Adam Morawiec

Orientations and Rotations

Springer

Berlin
Heidelberg
New York
Hong Kong
London
Milan
Paris
Tokyo

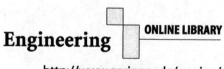

Engineering ONLINE LIBRARY

http://www.springer.de/engine/

Adam Morawiec

Orientations and Rotations

Computations in Crystallographic Textures

With 32 Figures

 Springer

Dr. Adam Morawiec
Polish Academy of Science
Institute of Metallurgy and Materials Science
Raymonta 25
30-059 Kraków
Poland

Cataloging-in-Publication Data applied for
Bibliographic information published by Die Deutsche Bibliothek
Die Deutsche Bibliothek lists this publication in the Deutsche Nationalbibliografie;
detailed bibliographic data is available in the Internet at <http://dnb.dd.de>

ISBN 3-540-40734-0 Springer-Verlag Berlin Heidelberg New York

Springer-Verlag Berlin Heidelberg New York
a member of BertelsmannSpringer Science + Business Media GmbH

http://www.springer.de

© Springer-Verlag Berlin Heidelberg 2004
Printed in Germany

Typesetting: Digital data supplied by author
Cover-Design: Erich Kirchner, Heidelberg
Printed on acid-free paper   62/3020 Rw 5 4 3 2 1 0

To my Wife.

Preface

THE ability to perceive orientations of objects is one of the basic characteristics of the space in which we live. Orientation changes are associated
with rotations of objects. Rotations, together with translations, constitute all
possible displacements of rigid objects. The concepts of orientation and rotation are of primary importance for science and engineering. In physics, these
notions appear in various contexts and at various scales: from the interpretation of angular momentum and molecular spectra in particle and atomic
physics, to the study of rigid body motion in classical mechanics, to the
analysis of rotations of celestial bodies in astronomy and considerations on
the global rotation of the Universe in cosmology. Our interest in orientations
comes from the prosaic area of crystallographic texture analysis. The main
subject of this branch of knowledge are orientations of crystallites constituting polycrystalline materials.

Most of basic results concerning rotations can be traced to L.Euler,
O.Rodrigues and A.Cayley. The related modern literature is quite extensive.
The description of rotations is the main subject of the book by Kuipers (1998).
There are numerous books on angular momentum containing chapters on rotations, e.g., Biedenharn (1965) or Altmann (1986). Also, many textbooks
on group theory use the rotation group as an example (e.g., Cornwell, 1984).
Moreover, works on kinematics of mechanisms (e.g., Bottema, 1979) or rigid
body motion (e.g., Whittaker 1904, Klein & Sommerfeld, 1910) can be quite
useful for someone trying to learn more about rotations. Refreshing views
on rotations can be found in books on related subjects (e.g., Coxeter's locus
classicus on regular polytopes (Coxeter, 1973)).

This book is primarily addressed to students and researchers working in
the area of crystallographic textures, especially those using numerical techniques, which are almost indispensable in texture analysis. Because of the
lack of introductory texts on computations in crystallographic textures, it is
necessary to cope with numerous publications, and thus, various styles and
conventions. The booklet is intended to be a handy tool for those writing
texture related computer programs. However, it is far from being just a col-
lection of useful formulas. I hope that the readers may actually learn from or
be inspired by these notes. The opening chapters (1 through 7), which contain
a relatively extensive introduction to rotations, can be useful for a very broad

audience which might include programmers coding computer animations, engineers working on robotic manipulators or spacecraft control, and researches analyzing limb or eye movements. The subjects covered in this part include parameterizations and geometry of $SO(3)$, orientation distributions and effects caused by symmetry. In subsequent chapters, some less general issues (the formal side of the determination of orientations from diffraction, effective properties) are considered. Representation theory of the rotation group – the topic considered at various levels in numerous books – is completely omitted. The aspects of the theory applied to crystallographic textures (series expansion) are extensively treated in Bunge's monography (1982).

As a non–mathematician, I emphasize material that aids computational automation while avoiding theorems and formal proofs. As for the notation, coordinates are deliberately used. This approach is less elegant but easier to apply than the more concise coordinate free notations. Moreover, readers must be aware of numerous conventions used in texts on rotations. An attempt was made to follow those frequently encountered in textures. (This led to awkward '−' signs in several expressions.) Besides the main path of material presentation, there is a second level marked by $\boxed{\boxdot}$. It contains auxiliary information of secondary importance for the considered subject (explanations, digressions, conventions, et cetera).

I would like to warn readers about two shortcomings of the text. The first one concerns the list of references. The reader must be aware that it is far from being complete. The list contains not necessarily the original or the most important contributions, but rather those which happened to be convenient for the author. The second warning concerns the style. I only hope that the 'international language of broken English' that I use, is precise enough so readers are not discouraged. Native speakers of English are asked for tolerance.

Finally, I would like to thank Helmut Schaeben (Institut für Geologie, TU Bergakademie Freiberg) for his comments on the manuscript. I carry sole responsibility for errors, and I shall be grateful if they are brought to my attention.

Kraków,

Adam Morawiec
June, 2003

Contents

1

Preliminaries

THE primary subject of crystallographic texture analysis are orientations of crystallites in a polycrystalline specimen. In order to be able to operate effectively on crystallite orientations, the understanding of general rules applicable to orientations of arbitrary finite objects is necessary. Loose thoughts of the first sections 1.1 — 1.4 are to recapitulate the relation between orientations and rotations and most essential attributes of the latter, and to give basic introduction to the composition of rotations and their matrix representation. Further on, slightly more advanced methods are applied. Throughout the chapter, a number of different ways of dealing with rotations are used: from classical geometry to purely algebraic methods. This is to show multiple faces of the issue and to introduce readers to some of the techniques employed further in the book.

1.1 Rotations as Displacements

Let us consider a finite three dimensional object and its orientation with respect to its environment. It is assumed here that both, the object and the environment are embedded in three dimensional Euclidean space. The concept of orientation is closely related to the notion of rotation. Object's rotation is seen either as a change of its orientation or as a relation between two (real or potential) orientations of the object. In the latter case, if one of the orientations is distinguished as a reference, all other orientations can be expressed in terms of rotations to that distinguished orientation.

Rotation is seen as a particular type of displacement in which a point – the center of rotation – is not displaced. It was noticed by Euler that a rotation about a point is equivalent to a rotation about a line. More precisely: Rotations about a point are displacements under which the location of that point remains unchanged, and rotations about a line are displacements under which points of the line retain their locations. According to the Euler's theorem, the configuration resulting from a rotation about a point can be achieved by a rotation about a line through that point.

Geometric proof

The Euler's theorem and all other theorems of this section can be easily proved by algebraic methods (see, e.g., Beatty, 1966). However, these calculations require appropriate apparatus which will be developed only later on. On the other hand, the older purely geometrical approach is based on very simple notions (and also has unquestionable charm). Therefore, basic results considered in this section will be justified in this traditional way.

The following proof of Euler's theorem can be found in the Whittaker's book (Whittaker, 1904). Let C be the center of rotation. Moreover, let A and B be two points of the object in its initial position. They are assumed to be not collinear with C. Positions of the points A and B after the rotation are denoted by A' and B', respectively. In the first step, two planes are constructed. The first one is perpendicular to the plane CAA' and bisects the angle ACA'; analogously, the second one is perpendicular to the plane CBB' and bisects the angle BCB'. If the two planes do not coincide, the line of their intersection is marked by L; see Fig. 1.1. Otherwise, L is the intersection of the planes CAB and $CA'B'$. Finally, if the planes CAB and $CA'B'$ coincide, L is the intersection between CAB and the bisecting planes. Now, it easy to see that in all cases the angles between L, CA and CB are equal to the angles between L, CA' and CB', respectively. Since the object is considered to be rigid, L retains its position when A and B are displaced to A' and B', respectively. Thus, L is the line of rotation. ⊠

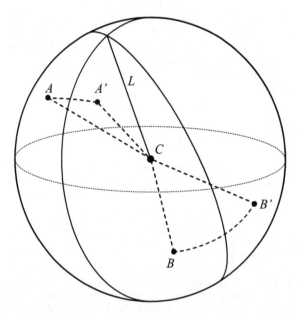

Fig. 1.1. Illustration of the first case in the proof of the Euler theorem for points A and B chosen to be at the same distance from C.

Obviously, the fixed line of rotation is not a complete determinant of the rotation. In order to characterize the latter uniquely, besides the line also the magnitude and sense of the rotation must be specified. The magnitude and the sense of the rotation can be established by the oriented dihedral angle between two planes, with the first plane determined by the fixed line and the initial location of a given point and the second one determined by the line, and the final location of that point; it is assumed that the point does not lie on the line. Alternatively, a rotation can be specified by a non–oriented angle and an oriented line – the *rotation axis*, and by setting the convention of, say right–handed, turns.

Another special type of a displacement is a translation. In a non–trivial translation, the line through initial and final locations of a point is parallel to analogous lines constructed for all points of the displaced object. The most general displacement in three dimensional Euclidean space can be expressed as a composition of a rotation and a translation. To see this, one can take an arbitrary base point and an oriented line segment from the point's original position to its final position after the displacement. The translation along that line by the distance equal to the length of the segment in the direction towards its end is the first element of the composition. The second is the displacement in which the final position of the base point is not moved, i.e., this is a rotation about that point.

Since the choice of the base point was arbitrary, a given general displacement can be accomplished by infinite number of different compositions of translations and rotations. However, for all these compositions, the rotation axes are parallel except for the null rotations, i.e., the identity transformation and complete–turn rotations causing no change in the configuration. Moreover, the rotation angle is unique up to the complete–turn rotations.

Evidence

In order to prove the above statements, let us take two different points A and B, and a non–trivial rotation transforming A on A' and B on B'. First, it will be shown that if the oriented line AB is parallel to the oriented line $A'B'$ and the lines have the same sense, then both are parallel to the rotation axis L: Analogously to the proof of the Euler's theorem, we take the bisecting planes of the angles ACA' and BCB', where C is a point on the rotation axis. The distances of A and A' from C are the same so the first bisector is perpendicular to AA'. Similarly, the second bisector is perpendicular to BB'. Since, by assumption, AB is parallel to $A'B'$, the lines AA' and BB' are parallel, too. Thus, the bisecting planes are parallel and must coincide because both contain the point C. Hence, the rotation axis is determined by the intersection of planes ABC and $A'B'C$. This leads to the axis parallel to AB. If the planes ABC and $A'B'C$ coincide, the axis is the intersection of that common plane and the common bisector of ACA' and BCB', and again the axis is parallel to AB.

Axis: Now, let the lines L_1 and L_2 be rotation axes in a displacement accomplished by two different compositions of translations and rotations. Take two different points A' and B' on L_2. A' and B' are obtained by the displacement of certain points A and B. The intermediate positions after translations are A_1, B_1 for the first composition, and $A_2 = A'$, $B_2 = B'$ for the second composition. See Fig. 1.2a. Since both pairs were obtained by translations of AB, the oriented line A_1B_1 is parallel to and has the same sense as the oriented line $A_2B_2 = A'B'$. A_1 and B_1 are transformed to A' and B' by the rotation about L_1 and the segments A_1B_1 and $A'B'$ are parallel. Thus, L_1 is parallel to $A'B'$ and to L_2.

Angle: To show uniqueness of the rotation angle take two different points A' and B' such that the line $A'B'$ intersects both rotation axes and is perpendicular to them. A' and B' are images of A and B in a displacement accomplished by two different compositions of translations and rotations. The intermediate positions after

translations are A_1 and B_1 for the first composition, and A_2 and B_2 for the second composition. See Fig. 1.2b. The line A_1B_1 is parallel to A_2B_2 because both were obtained by translations of the same segment AB. The line A_1B_1 intersects $A'B'$ at the first rotation axis because $A'B'$ is obtained by rotation of A_1B_1 by the axis perpendicular to $A'B'$. Analogously, the line A_2B_2 intersects $A'B'$ at the second rotation axis. Thus, A_1B_1 and A_2B_2 are two parallel lines intersected by the line $A'B'$. Hence, the corresponding angles are equal and the rotation angles are equal.
⊠

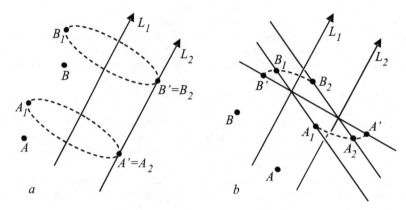

Fig. 1.2. Sketches for the proof of parallelism of rotation axes (a) and uniqueness of rotation angle (b).

◻ *Order of translation and rotation*
The property proved above allows us to deduce that the following two displacements:
– the first one composed of a translation followed by a rotation and
– the second one composed of the same rotation and the same translation
(i.e., with the order of the operations switched) differ by a translation. ⊠

◻ *Chasles theorem*
Possible compositions of translations and rotations corresponding to a given displacement are limited by the Chasles' theorem: the most general displacement in three dimensional Euclidean space can be represented by a so–called screw, i.e., a composition of a rotation and a translation *along the direction of the rotation axis*. The screw is unique, i.e., different screws representing one and the same displacement differ by null rotations. ⊠

When only the object's orientation is considered, translations are disregarded. The equivalence class of all displacements which differ by a translation represents one orientation. If a universal reference orientation is distinguished, a given orientation of the object is determined by the rotation from the reference orientation to that given orientation. There is a one–to–one correspondence of object's orientations and such rotations. However, as was already mentioned, the two notions are usually distinguished; a rotation is seen as a process whereas an orientation as a state of an object. E.g., after a full pirouette (complete–turn), initial and final orientations are equivalent but the pirouette as a rotation process is not equivalent to 'no motion'. (Altmann (1986) uses the terms 'rotation' and 'spinning' for the state and the process, respectively.) In this text, more often than not, 'rotation' is identified with 'orientation'. In rare places where the 'rotation' corresponds to a displacement with its history, the actual meaning is indicated by context.

1.2 Composition of Rotations

Two successive rotations about the same fixed point do not change the location of that point and thus, their composition is a rotation about the point. The question is what is the relation of the net rotation to the composing rotations. To give an answer to that question, it is convenient to use a spherical triangle on the sphere centered at the fixed point. Let the vertices and the vertex angles of the triangle be denoted by the same letters A, B and C. Moreover the center of the sphere will be denoted by Z.

By the theorem of Rodrigues and Hamilton, three successive rotations
– about the axes AZ, BZ and CZ,
– through the angles $2A$, $2B$ and $2C$, respectively,
– with the sense of the rotations opposite to the one indicated by the order of the vertices ABC
produce no net displacement, i.e., they restore the rotated object to its original position.

Explanation
It is easy to see the correctness of the theorem by considering positions of the triangle ABC after each of the three rotations. Fig. 1.3 is a planar sketch of the configuration on the sphere. Analysis of angles in the displayed triangles proves that the final position overlaps the initial one. ⊠

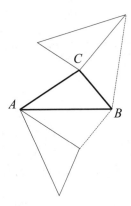

Fig. 1.3. Planar sketch of the spherical triangles used in the proof of the Rodrigues–Hamilton theorem.

The Rodrigues–Hamilton theorem allows the result of the composition of rotations to be determined: The composition of the first two rotations of the Rodrigues–Hamilton theorem is equal to the rotation which differs from the third one by the sense of rotation.

There is another well known classical theorem (of Donkin) similar to that of Rodrigues and Hamilton. Let the side angles of the spherical triangle considered above be denoted by AB, BC and CA. Moreover, let P_{AB}, P_{BC} and P_{CA} denote lines through Z and perpendicular to the planes ZAB, ZBC and ZCA, respectively. By the Donkin's theorem, three successive rotations
– about the axes P_{AB}, P_{BC} and P_{CA},
– through the angles $2\,AB$, $2\,BC$ and $2\,CA$, respectively,
– with the sense of rotations indicated by the order of the vertices ABC
produce no net displacement, i.e., they restore the rotated object to its original position.

□ *More on Donkin's theorem*
This can be proved by considering positions of a spherical triangle after subsequent rotations. Fig. 1.4 is a planar sketch of the configuration on the sphere: the position of the bottom left triangle is restored by the rotations described in the Donkin's theorem.

The Donkin's theorem for a given triangle is *de facto* the Rodrigues–Hamilton theorem applied to the spherical triangle polar to the given one, and *vice versa*. (See, e.g., Berger, 1987, for the definition of the polar triangle.) One theorem follows from the other by the application of relations between angles of mutually polar triangles.

The Donkin's theorem can be generalized to a spherical polygon $ABCD...E$: successive rotations about axes P_{AB}, P_{BC}, P_{CD}, ..., P_{EA} through the angles $2\,AB$, $2\,BC$, $2\,CD$, ..., $2\,EA$ with the sense of rotation indicated by the order of the vertices $ABCD...E$ produce no net displacement. This is already visible from the construction used for proving the triangle case. Alternatively, the triangles ABC, ACD, ... can be considered. With the added side AC traversed twice in subsequent rotations with opposite senses, the net addition gives no displacement. On the other hand, by the original Donkin's theorem, there is no displacement in the case of an individual triangle. Analogous generalization can be formulated for the Rodrigues–Hamilton theorem. ⊠

Like Rodrigues–Hamilton theorem, the Donkin's theorem allows us to determine the composition of rotations: The composition of the first two rotations of the Donkin's theorem is equal to the third rotation but with the opposite sense of rotation.

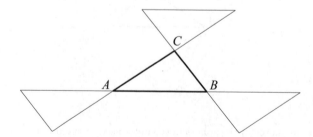

Fig. 1.4. Planar sketch for the proof of the Donkin's theorem.

The rotation by the angle of measure zero composed with a given rotation (in arbitrary order) has the same effect as the latter; the former is the identity transformation. Moreover, two successive rotations about the same line by a given angle but in opposite directions do not change the object's orientation. These are mutually inverse rotations. By counter–example, one can show that, in general, rotations do not commute. It is trivial to notice that coaxial rotations do commute and the null rotations commute with all rotations. Moreover, two half–turns (i.e., rotations by half of the full circle) about perpendicular axes commute.

1.3 Improper Rotations

In crystallography, not just displacements but a broader class of distance preserving (isometric) transformations must be considered. Chiral crystals exist in two different (enantiomorphic) forms with structures related by a reflection. Therefore, it is assumed that an object can be not only displaced but also converted into its mirror image. Isometric transformations having a fixed point and involving reflections are called improper rotations (or rotoinversions). A most general improper rotation can be seen as a composition of a reflection and a proper rotation. Since handedness of an object is preserved by proper rotations and is changed in reflections, an improper rotation changes handedness.

A rotation in which every line segment from a point to its image is bisected by the center of rotation is called inversion. In the three dimensional space, the inversion is an improper rotation. It can be seen as a composition of a half–turn about a given axis and the reflection with respect to the plane perpendicular to that axis. As the inversion is unique, the choice of the rotation axis is arbitrary.

Since the inversion changes handedness, an improper rotation can be achieved by a composition of a proper rotation and the inversion. For each proper rotation there is an invariant plane perpendicular to the rotation axis; an arbitrary point of such a plane is transformed to a point of the same plane. In the inversion, each plane through the center is invariant. Therefore, an improper rotation has an invariant plane. The invariant plane of an improper rotation is unique unless the improper rotation is the inversion. The decomposition of an improper rotation into a proper rotation and the inversion is unique. The order of a proper rotation and the inversion in the decomposition is immaterial.

Other decompositions of similar kind may be considered:

1. An improper rotation can be achieved by a composition of a proper rotation and reflection with respect to the plane perpendicular to the rotation axis. The plane in this decomposition is the invariant plane of the improper rotation. The order of operations in the decomposition is immaterial. Moreover, it is easy to deduce that the decomposition is unique unless the improper rotation is the inversion.

2. It is also noteworthy that the result of a proper rotation can be achieved by a composition of two mirror transformations. With the mirror planes containing the axis of the proper rotation, the existence of the composition equivalent to the rotation can be shown by a simple geometrical construction in

the plane perpendicular to the axis. It is also clear from the construction that the mirror planes are not unique.

3. Finally, it follows from the previous points that an arbitrary improper rotation can be decomposed into three reflections. In particular, the inversion is equivalent to the composition of reflections with respect to three perpendicular planes.

In large parts of this booklet, we are dealing with proper rotations. Wherever improper rotations are allowed, this is explicitly declared.

1.4 Matrix Representation

We started with a consideration of an orientation of a three dimensional object in the three dimensional Euclidean space. The object, however, may have a lower dimension. E.g. in the case of nematic liquid crystals, long stick–like molecules with the symmetry of the ellipsoid of revolution are subject of investigation. Their orientations are determined by orientations of the axes of revolution, i.e., by orientations of one–dimensional objects. In such cases, a more general approach with p-dimensional objects embedded in N–dimensional ($N \geq p$) Euclidean space would be applicable.

In order to perform calculations concerning orientations, they must be represented somehow by numbers. An orientation can be determined by providing coordinates of some vectors attached to the object. Let the coordinates be specified in a coordinate system with the origin at the center of rotation. The number of vectors is related to the dimension of the object. For a p–dimensional object it is enough to select p linearly independent vectors $r^{(i)}$, ($i = 1, .., p$). If the space is N–dimensional, the orientation can be expressed as an $p \times N$ matrix O with the selected vectors as its rows ($O_{ij} = r_j^{(i)}$). Without loosing generality of further considerations, we assume that the vectors are of unit magnitude and mutually orthogonal, and the coordinate system is Cartesian. Thus, the matrix O satisfies the condition

$$OO^T = I_p , \tag{1.1}$$

where I_p is $p \times p$ identity matrix. The topological space with $p \times N$ matrices satisfying (1.1) as points of the space is a base of the so–called Stiefel manifold. The dimension of the manifold is $Np - p(p+1)/2$. Hence, for $N = p = 3$, the dimension is three.

> *Number of independent parameters*
> The first vector of the set is a unit vector and $N - 1$ parameters are needed to specify it. The second vector is unit and perpendicular to the first one, so $N - 2$ parameters are needed. The third vector is unit and perpendicular to the first and the second and requires $N - 3$ parameters. In general the number of parameters for all p vectors is $\sum_{i=k}^{p}(N - k) = Np - p(p+1)/2$. $\boxtimes$

In the particular case when the dimension of the object is equal to the dimension of the space ($p = N$) the matrices are square; with the condition (1.1) satisfied, they are called orthogonal. The orthogonal matrices are invertible because $(\det(O))^2 = \det(O^T O) = \det(I_N) = 1$; the inverse O^{-1} of O is given by $O^{-1} = O^T$. Thus, besides $OO^T = I_N$ also $O^T O = I_N$ occurs.

The next aspect to be considered is the relationship between matrices corresponding to two orientations. For two orientations of an object, two sets of vectors $r^{(i)}$ and $r'^{(i)}$, $(i = 1, .., p)$ must be specified. The corresponding matrices are $O_{ij} = r_j^{(i)}$ and $O'_{ij} = r_j'^{(i)}$. Mutual relationship between the two orientations can be expressed via scalar products of vectors $r^{(i)}$ and $r'^{(i)}$. Let R be the $p \times p$ matrix containing these products, i.e. $R_{ij} = r'^{(i)} \cdot r^{(j)}$. This can be rewritten using the matrices O and O'

$$R = O'O^T . \qquad (1.2)$$

The matrix R relates the two orientations and is independent of the choice of the initial reference frame used for defining O and O'.

Let us concentrate now on the case $p = N$ with orthogonal matrices O and O'. The matrix R is also orthogonal because $RR^T = O'O^T(O'O^T)^T = O'O^TOO'^T = I_N$. Taking (1.2) in the form $O' = RO$, R can be interpreted as a transformation from the orientation O to O'; in other words, the orthogonal matrix R determines a (proper or improper) rotation of the N dimensional object from the orientation corresponding to O, to the orientation corresponding to O'. If the two sets of vectors $r^{(i)}$ and $r'^{(i)}$ are linked with two different objects, R is sometimes referred to as *misorientation*.

What is the determinant of R? It is related to determinants of O and O' through $\det(R) = \det(O)\det(O')$. Thus, if orientations of an object are allowed to correspond to orthogonal matrices with both positive and negative determinants, then $\det(R) = \pm 1$. In three dimensions, it means that the object is allowed to change handedness. If the change of that type is forbidden and orientations are determined by orthogonal matrices with determinants having the same sign, then the rotations correspond to orthogonal matrices with determinants equal to $+1$. Such matrices are called special orthogonal.

There is a one–to–one correspondence between orthogonal matrices and general (i.e., proper and improper) rotations. Similarly, there is a one–to–one correspondence between special orthogonal matrices and proper rotations. The (special) orthogonal matrices are said to represent (proper) rotations. Therefore, frequently, a shortcut is used and no distinction is made between these corresponding notions.

As was already mentioned, orientations can be uniquely ascribed to rotations. This is achieved by the assumption that there is a special reference orientation. If $O = I$ is taken as this reference orientation, then (1.2) takes the form $R = O'$, i.e., a given orientation and the corresponding rotation are represented by the same matrix.

Now, if O'' is another orientation related to O' by the rotation $R' = O''O'^T$, then the rotation relating O'' and O is given by the product of matrices R' and R. It follows from $O'' = R'O' = R'RO$. Clearly, the composition of rotations is represented by the multiplication of corresponding matrices. The composition of rotations is associative, which means that the rotation $(R'R)$ followed by R'' has the same effect as R followed by $(R''R')$, i.e., $R''(R'R) = (R''R')R$. There exist the 'identity rotation' (represented by the identity matrix I) which does not change orientation of an object and satisfies $IR = RI = R$ for an arbitrary rotation R. Finally, each rotation has its inverse. From the viewpoint of algebra, the set of rotations with the composition as the operation on them is a group. The corresponding (isomorphic) group of $N \times N$ orthogonal matrices is denoted by $O(N)$.

Moreover, the set limited to proper rotations constitutes a group, too. To see that it is enough to notice that proper rotations are closed under composition and taking inverses. Corresponding $N \times N$ special orthogonal matrices constitute the group called in mathematics $SO(N)$.

If a vector x is given in the form of linear combinations $x = x^j r^{(j)} = x'^i r'^{(i)}$, one has $x'^i = x \cdot r'^{(i)} = x^j r^{(j)} \cdot r'^{(i)}$ or

$$x'^i = R_{ij} x^j \ , \tag{1.3}$$

which is a formula for transforming vector coordinates.

With rotations perceived as displacements of an object with respect to a reference frame, there are actually two ways of realizing and expressing the change. In the case of so–called active rotations, the object (e.g., a vector) is displaced and the frame is kept fixed. Alternatively, if the object is considered to be fixed and there are variable frames, the rotation is referred to as passive.[1] The passive convention is used in crystallographic textures.

1.5 Formal Approach to Orthogonal Transformations

Rotations can be seen as transformations of vectors. Since the object of a rotation remains rigid, the mutual configuration of vectors undergoing such transformation remains unchanged. In particular the scalar product of vectors is invariant, i.e., the scalar product of two vectors is equal to the scalar product of their images obtained by the rotation transformation. It is worth to consider the issue from a more formal point of view. Transformations vector spaces which in a way generalize rotations are an elementary subject of algebra. Let X be real vector space of dimension N and g a symmetric, positive definite bilinear form, i.e., g is linear in each of its arguments and

$$g(x, y) = g(y, x) \ \text{ and } \ g(x, x) > 0 \text{ for } x \neq 0 \ .$$

That form is a generalization of the scalar product. By definition, operator $O : X \to X$ is orthogonal if

$$g(Ox, Oy) = g(x, y) \tag{1.4}$$

for x, y in X. Orthogonal operators are linear: it can be easily shown that with α being a real number, $g(O(\alpha y+z)-(\alpha Oy+Oz), O(\alpha y+z)-(\alpha Oy+Oz)) = 0$, and hence, because of the positive definiteness of g,

$$O(\alpha y + z) = \alpha Oy + Oz \ .$$

---

[1] With a vector undergoing an active rotation described by (1.3), x^i and x'^i are coordinates of two vectors in the same coordinate system; in the passive interpretation, x^i and x'^i are seen as coordinates of the same vector in two coordinate systems.

The norm $\|x\|_g = \sqrt{g(x,x)}$ is invariant in orthogonal transformations, and the metric $d(x,y) = \|x - y\|_g$ satisfies $d(Ox, Oy) = d(x,y)$, i.e., orthogonal transformations are isometric.

The operator O is injective, i.e., if $Ox = Oy$, then $x = y$. This statement follows from the sequence $g((x - y), (x - y)) = g(O(x - y), O(x - y)) = g((Ox - Oy), (Ox - Oy)) = g(0,0) = 0$ leading to $x - y = 0$. The composition of orthogonal operators O' and O'' gives an orthogonal operator $O''O'$: $g(O''O'x, O''O'y) = g(O'x, O'y) = g(x,y)$. Also, the identity operator I given by $Ix = x$ (for each x in X) is orthogonal and $IO = O = OI$ for an arbitrary orthogonal operator O. Since O is linear and injective, for every x in X exists x' in X such that $Ox' = x$. Thus, there exists an inverse orthogonal operator O^{-1} such that $O^{-1}O = I = OO^{-1}$. Hence, orthogonal operators constitute a group.

It is easy to verify that the operator O_p defined by

$$O_p(x) = x - \frac{2g(p,x)}{g(p,p)}p \,, \quad \text{where } p(\neq 0) \text{ and } x \text{ are in } X \,, \tag{1.5}$$

is orthogonal. The transformation by O_p is called reflection with respect to the hyperplane orthogonal to p. According to Cartan theorem, the group of orthogonal transformations is generated by reflections; an arbitrary orthogonal operator O can be expressed as a composition of (at most $2N$) reflections $O = O_{p_k}...O_{p_2}O_{p_1}$, $k \leq 2N$. Transformations expressible by *even* number of reflections constitute a subgroup of the group of orthogonal transformations and are called *special* orthogonal.

Proof
The Cartan theorem is proved by induction on the dimension of X. It is obvious for $N = 1$. Let it be true for the space of dimension $N - 1$. For a given non–zero vector x and an orthogonal transformation O of the N dimensional space X, one has either $Ox = x$ or $Ox - x \neq 0$.

In the first case, O restricted to the hyperplane orthogonal to x is an orthogonal transformation and can be expressed as a composition of reflections $O'_{p_k}...O'_{p_2}O'_{p_1}$ since the dimension of the hyperplane is $N - 1$. A reflection O'_p in the hyperplane orthogonal to x can be extended to X by $O_p(ax + y) = ax + O'_p(y)$, where y is in the hyperplane. Thus, O is a composition of the extensions $O_{p_k}...O_{p_2}O_{p_1}$.

The second case $(Ox - x \neq 0)$, will be reduced to the first one. Let us take the reflection of Ox with respect to the hyperplane orthogonal to $q = x - Ox$

$$O_q(Ox) = Ox - \frac{2g(Ox, x - Ox)}{g(x - Ox, x - Ox)}(x - Ox) \,.$$

Since $g(x - Ox, x - Ox) = 2g(x,x) - 2g(Ox,x) = -2g(Ox, x - Ox)$, there occurs

$$O_q O(x) = O_q(Ox) = Ox + (x - Ox) = x \,.$$

From the previous case, $O_q O$ can be expressed as a composition of reflections: $O_q O = O_{p_k}...O_{p_2}O_{p_1}$. Since $O_q O_q O = O$, there occurs $O = O_q O_{p_k}...O_{p_2}O_{p_1}$. $\boxtimes$

Possible generalizations
Further generalizations of orthogonal operators are obtained by abandoning the requirement of the positive definiteness of g or by allowing the complex vector space. $\boxtimes$

1.5.1 Eigenvalues and eigenvectors of orthogonal operators

Even in the general framework of the previous section, some spectral analysis can be performed. Let us assume for a while that λ is real, ξ is a non–zero element of X and

$$O\xi = \lambda\xi \ . \tag{1.6}$$

Thus, from (1.4) follows

$$(1-\lambda^2)g(\xi,\xi) = g(\xi,\xi)-g(\lambda\xi,\lambda\xi) = g(\xi,\xi)-g(O\xi,O\xi) = g(\xi,\xi)-g(\xi,\xi) = 0 \ . \tag{1.7}$$

Because of positive definiteness of g, one has $\lambda = \pm 1$, i.e., if there is a real eigenvalue, it is equal either to 1 or to -1.

The equation (1.6) may have complex solutions. Let λ be a complex number and ξ be element of the complexification of X, i.e., $\lambda = a + bi$, $\xi = x + iy$, where a and b are real, x, y are in X and $i^2 = -1$. The multiplication $\lambda\xi$ is defined by $\lambda\xi = (ax - by) + i(ay + bx)$ and the vector conjugate to ξ is $\bar{\xi} = x - iy$. For the arguments $x+iy$, $s+it$ (s,t in X), the form g is determined by $g(x + iy, s + it) = g(x,s) + ig(x,t) + ig(y,s) - g(y,t)$. Moreover, for $O\xi$, we take $O\xi = O(x + iy) = Ox + iOy$. Based on these definitions,

$$\overline{O\xi} = \overline{Ox + iOy} = Ox - iOy = O(x - iy) = O\bar{\xi} \ ,$$

and there occurs $(1 - \lambda\bar{\lambda})g(\xi,\bar{\xi}) = g(\xi,\bar{\xi}) - g(\lambda\xi,\overline{\lambda\xi}) = g(\xi,\bar{\xi}) - g(O\xi,\overline{O\xi}) = g(\xi,\bar{\xi}) - g(O\xi,O\bar{\xi}) = 0$. Since $g(\xi,\bar{\xi}) = g(x+iy, x-iy) = g(x,x)+g(y,y) > 0$, we have $\lambda\bar{\lambda} = 1$ or $\lambda = e^{i\phi}$, where ϕ is a real number. If λ and ξ are an eigenvalue and the corresponding eigenvector of O, then so are $\bar{\lambda}$ and $\bar{\xi}$

$$O\bar{\xi} = \overline{O\xi} = \overline{\lambda\xi} = \bar{\lambda}\bar{\xi} \ .$$

Therefore, if the dimension of the vector space is odd, each orthogonal operator has at least one real eigenvalue (1 or -1). In other words, in odd dimensional spaces there is a one dimensional subspace invariant under the orthogonal operation. In three dimensional space, for O with $\det O = +1$, it is the fixed line of the Euler theorem.

Assuming that vectors $\{e_i, i = 1, ..., N\}$ constitute a basis of X, Oe_i can be expressed as

$$Oe_i = O^j{}_i e_j \ .$$

With $g_{ij} = g(e_i, e_j)$, there occurs $O^k{}_i O^l{}_j g_{kl} = O^k{}_i O^l{}_j g(e_k, e_l) = g(Oe_i, Oe_j) = g(e_i, e_j) = g_{ij}$. In a more compact notation

$$O^T g O = g \ . \tag{1.8}$$

Assuming that the basis is orthonormal, i.e., with $g = I$, there occurs $O^T O = I$, i.e., a particular form of (1.1).

Let an eigenvector corresponding to a real eigenvalue of O be chosen as e_K. The relations $Oe_K = O^i{}_K e_i$ and $Oe_K = \pm e_K$ lead to $O^i{}_K = \pm\delta^i_K$. Let the lower (upper) index enumerate columns (rows) of the corresponding matrix. Thus, the K-th column contains only zeros except K-th entry which equals

1 or -1. Moreover, other entries of the matrix are bounded by conditions $g_{Kj} = O^k{}_K O^l{}_j g_{kl} = \pm \delta^k_K O^l{}_j g_{kl} = \pm O^l{}_j g_{Kl}$. In the orthonormal basis ($g_{ij} = \delta_{ij}$), this leads to $O^K_j = \pm \delta_{Kj}$. Thus, in the case of orthogonal matrix, also the K-th row contains zeros everywhere except the K-th entry. The matrix obtained by deleting K-th row and K-th column is orthogonal in the subspace of X perpendicular to the vector e_K and the construction can be repeated if there are any other real eigenvalues.

Let us consider now the case of complex eigenvalues. For $\xi = x + iy$, where x, y are elements of X and $\lambda = e^{i\phi} = \cos\phi + i\sin\phi$, the equation $O\xi = \lambda\xi$ gives

$$Ox + iOy = (\cos\phi + i\sin\phi)(x + iy) = (\cos\phi\, x - \sin\phi\, y) + i(\sin\phi\, x + \cos\phi\, y) \ .$$

Thus,

$$Ox = x\cos\phi - y\sin\phi \quad \text{and} \quad Oy = x\sin\phi + y\cos\phi \ . \qquad (1.9)$$

To a complex eigenvalue of an orthogonal operator corresponds a two dimensional subspace invariant under the orthogonal operation. Since $(1 - \lambda^2)g(\xi, \xi) = 0$, $\lambda^2 = e^{2i\phi}$, and λ is not real, one has $g(\xi, \xi) = 0$. On the other hand, there occurs

$$g(\xi, \xi) = g(x + iy, x + iy) = g(x, x) - g(y, y) + 2ig(x, y) \ .$$

Thus, $g(x, x) = g(y, y)$ and vectors x and y are orthogonal ($g(x, y) = 0$). If these vectors are chosen as basis vectors $e_K = x$ and $e_L = y$, due to Eq.(1.9),

$$O^K{}_K = \cos\phi, \quad O^L{}_K = -\sin\phi, \quad \text{and} \quad O^i{}_K = 0 \text{ for } i \neq K, L \ ,$$

$$O^K{}_L = \sin\phi, \quad O^L{}_L = \cos\phi, \quad \text{and} \quad O^i{}_L = 0 \text{ for } i \neq K, L \ .$$

Thus, in K-th and L-th columns, only K-th and L-th row entries are non–zero. The same concerns rows if the basis is orthonormal (and the matrix is orthogonal). The matrix obtained by deleting K-th row and K-th column and L-th row and L-th column is orthogonal in the subspace of X perpendicular to the vectors e_K and e_L and the construction can be repeated.

The above analysis shows that rotations 'take place' in 2 dimensional planes. This observation is crucial for understanding rotations in spaces of high dimension. The rotation axis is a concept applicable only in three dimensional space. As this particular case is of interest here, rotation axes will be frequently used.

In the three dimensional space, each orthogonal operator has at least one real eigenvalue with absolute value of 1. If there is more than one such eigenvalue, all three eigenvalues have absolute values of 1 because complex eigenvalues appear in pairs. The decomposition of O described in previous paragraphs takes one of the forms

$$\begin{bmatrix} \cos\varphi & \sin\varphi & 0 \\ -\sin\varphi & \cos\varphi & 0 \\ 0 & 0 & \pm 1 \end{bmatrix} , \quad \begin{bmatrix} \cos\varphi & 0 & -\sin\varphi \\ 0 & \pm 1 & 0 \\ \sin\varphi & 0 & \cos\varphi \end{bmatrix} , \quad \begin{bmatrix} \pm 1 & 0 & 0 \\ 0 & \cos\varphi & \sin\varphi \\ 0 & -\sin\varphi & \cos\varphi \end{bmatrix} .$$

The new basis e'_i, ($i = 1, 2, 3$) defined as $e'_j = O^i{}_j e_i$ has the same handedness as e_i, ($i = 1, 2, 3$) only if $\det(O) > 0$. In that case, the orthogonal transformation corresponds to a proper rotation. Otherwise ($\det(O) < 0$), it represents an improper rotation. From the above considerations follows that each proper rotation has at least one eigenvalue of 1 and each reflection has at least one eigenvalue of -1. The eigenvector corresponding to the real eigenvalue has a direct connection to the rotation axis of the Euler theorem; the vector determines the direction of the axis.

Unitary operators

For the introduction of Cayley–Klein parameters the so–called unitary matrices will be used. As there is an analogy between orthogonal and unitary operators, a brief information on the latter is given below.

Let X be a vector space over the field of complex numbers, and let f be a positive definite hermitean form; i.e., f is assumed to be antilinear in the first, and linear in the second arguments, and

$$f(x, y) = \overline{f(y, x)} \text{ and } f(x, x) > 0 \text{ for non-zero } x .$$

Operator $U : X \to X$ is unitary if

$$f(Ux, Uy) = f(x, y)$$

for x, y in X. Unitary operators are linear. The norm $\|x\|_f = \sqrt{f(x, x)}$ is invariant in unitary transformations and with the metric $d_f(x, y) = \|x - y\|_f$, the unitary operation is isometric.

Assuming that vectors $\{e_i, i = 1, ..., N\}$ constitute a basis, Ue_i can be expressed as

$$Ue_i = U^j{}_i e_j .$$

With $f_{ij} = f(e_i, e_j)$, there occurs

$$\overline{U^k{}_i} U^l{}_j f_{kl} = \overline{U^k{}_i} U^l{}_j f(e_k, e_l) = f(Ue_i, Ue_j) = f(e_i, e_j) = f_{ij} .$$

In a more compact form $U^\dagger f U = f$, where $(U^\dagger)^i{}_j = \overline{U^j{}_i}$. Hence, $\det(U^\dagger) \det(f) \det(U) = \det(f)$ and thus $\overline{\det(U)} \det(U) = 1$ or

$$|\det(U)| = 1 .$$

Assuming that the basis is orthonormal, i.e., with $f_{ij} = \delta_{ij}$, there occurs

$$\overline{U^k{}_i} U^k{}_j = \overline{U^k{}_i} U^l{}_j \delta_{kl} = \delta_{ij} ; \tag{1.10}$$

such a matrix U is called unitary. Unitary matrices satisfying $\det(U) = 1$ are referred to as special unitary. The group they constitute is denoted by $SU(N)$. Based on Eq.(1.10), the entries a, b, c, d of a $SU(2)$–type matrix

$$\begin{bmatrix} a & b \\ c & d \end{bmatrix}$$

satisfy the conditions $ad - bc = 1$ and

$$d = \overline{a} , \qquad c = -\overline{b} . \tag{1.11}$$

There is a close relationship between $SU(2)$ and $SO(3)$. It will be discussed in chapters 2 and 3. ⊠

1.5.2 Formalism of Clifford Algebras

Precise definition of Clifford algebras is rather involved. Therefore, we will limit this section to a brief informal description of the algebras. An algebra can be constructed from a set of elements of a linear space which has a product operation, with products of the elements being elements of the set. Let X_1 denote a linear space of dimension N over the field of real numbers. Let g be a non–degenerated symmetric bilinear form on X_1. The Clifford product, by definition, commutes with the scalar multiplication, is associative, distributive under addition, and for x belonging to X_1 satisfies

$$x \diamond x + g(x, x) = 0 \ . \tag{1.12}$$

Let v_i, $i = 1, ..., N$, be a basis of X_1. Let the linear space X_m $(m = 1, ..., N)$ be spanned by the basis $v_{i_1} \diamond ... \diamond v_{i_m}$, where $1 \leq i_1 < ... < i_m \leq N$. X_0 is a field of real numbers. The Clifford algebra is the direct sum of the X_m spaces $(m = 0, ..., N)$ with the Clifford product. Depending on the dimension of X_1 and signature of g, various Clifford algebras arise.[2]

> **Example**
> Let two dimensional X_1 be spanned by vectors b_1 and b_2. The bases of X_0 and X_2 are $\{1\}$ and $\{b_1 \diamond b_2\}$, respectively. Thus the linear space of the algebra is spanned by
> $$\{1, b_1, b_2, b_1 \diamond b_2\} \ .$$
> Let the metric be given by $g(b_i, b_j) = -\delta_{ij}$. Because of (1.12), the vectors b_i satisfy
> $$b_1 \diamond b_1 = 1 = b_2 \diamond b_2 \quad \text{and} \quad b_1 \diamond b_2 = -b_2 \diamond b_1 \ . \tag{1.13}$$
> Let us calculate the product $(1 + b_1 \diamond b_2) \diamond (1 - b_1 \diamond b_2)/2$. Since the Clifford product is distributive,
> $$(1 + b_1 \diamond b_2) \diamond (1 - b_1 \diamond b_2) = 1 + b_1 \diamond b_2 - b_1 \diamond b_2 - b_1 \diamond b_2 \diamond b_1 \diamond b_2 \ .$$
> Moreover, because of $b_1 \diamond b_2 \diamond b_1 \diamond b_2 = -b_1 \diamond b_1 \diamond b_2 \diamond b_2 = -1$, we have
> $$(1/2)(1 + b_1 \diamond b_2) \diamond (1 - b_1 \diamond b_2) = 1 \ . \tag{1.14}$$
> Similarily, the Clifford products of $p = b_2 + b_1 \diamond b_2$ and $q = 1 - b_1$ are
> $$p \diamond q = 2b_2 + 2b_1 \diamond b_2 \ , \quad q \diamond p = 0 \ . \tag{1.15}$$
> These examples illustrate some properties of the Clifford product. ⊠

Based on (1.15), it is clear that the Clifford product is neither commutative nor anti–commutative. The second relation of (1.15) exemplifies the existence of zero divisors. This means that there are non–zero elements having no inverse with respect to Clifford product. On the other hand, (1.14) is the product of invertible elements. It is easy to see that the direct sum of the spaces X_m with even m with the Clifford product constitutes an (even) subalgebra of $\mathcal{C}_N$. It is denoted by $\mathcal{C}_N^+$.

---

[2] However, the variety is, in a sense, limited since higher dimensional Clifford algebras are related to those of lower dimension by the so–called periodicity theorem.

Since $2g(x,y) = g(x,x) + g(y,y) - g(x-y,x-y) = -x \diamond x - y \diamond y + (x-y) \diamond (x-y)$, then

$$g(x,y) = -(x \diamond y + y \diamond x)/2 . \tag{1.16}$$

This relationship will be used below.

Grassmann algebra
With $x \wedge y := (x \diamond y - y \diamond x)/2$, there occurs

$$x \diamond y = x \wedge y - g(x,y)$$

and the product $\wedge$ is anti–commutative. The direct sum of X_m $(m = 0, ..., N)$ with associative, distributive and anti–commutative exterior product $\wedge$ is the more familiar Grassmann algebra. When g is set to be identically zero, then the Clifford algebra becomes the Grassmann algebra. ⊠

Isomorphisms
Although, we would like to avoid unnecessary multiplication of notions, some are difficult to substitute. For instance, it is convenient to understand the concepts of homo– and isomorphism. Let us give definitions of these terms for groups.

Homomorphism of groups G and G' is a mapping, say ψ, of G in G' such that

$$\psi(AB) = \psi(A)\psi(B)$$

for all elements A and B of G. If the homomorphism is *onto* and *one–to–one* the groups are said to be isomorphic. 'Isomorphic' means that the groups are structurally identical.

In a similar way, isomorphism is defined for other algebraic structures. One of them is isomorphism of algebras. Also here, 'isomorphic' means that there is a full correspondence of elements and operations and that the algebras are identical from the formal viewpoint. For instance, the algebra given in the example above can be identified with the algebra of real 2×2 matrices. This can be verified by considering the mapping

$$1 \rightarrow I_2 , \quad b_1 \rightarrow \begin{bmatrix} 0 & 1 \\ 1 & 0 \end{bmatrix} =: \sigma_1 , \quad b_2 \rightarrow \begin{bmatrix} 1 & 0 \\ 0 & -1 \end{bmatrix} =: \sigma_3 .$$

The matrices σ_1 and σ_3 satisfy $\sigma_1 \sigma_1 = I_2 = \sigma_3 \sigma_3$ and $\sigma_1 \sigma_3 = -\sigma_3 \sigma_1$ i.e., relations analogous to (1.13). ⊠

Exploration of Clifford algebras in full complexity is beyond the scope of these notes. We only want to emphasize their relation to orthogonal transformations of Euclidean space. With such a narrow purpose, it is enough to limit further considerations to positive definite g and this will be assumed from now on. In this simple case, one can take an orthonormal basis b_i $(i = 1, ..., N)$, i.e., such that $g(b_i,b_j) = \delta_{ij}$. From (1.16) follows $b_i \diamond b_j + b_j \diamond b_i = -2\delta_{ij}$. With positive definite g, the chain of Clifford algebras begins with well known entities:
• C_0 of dimension 1 corresponds to the real numbers. The even subalgebra C_0^+ is empty.
• C_1 of dimension 2 has the basis $\{1, b_1\}$. The general form of its element is $c + db_1$ with $b_1 \diamond b_1 = -1$. Thus, C_1 has the same structure as (i.e., is isomorphic to) the algebra of the complex numbers. The even subalgebra C_1^+ corresponds

to the real numbers.

- C_2 of dimension 4 has the basis $\{1, b_1, b_2, b_1 \diamond b_2\}$. The multiplication table (with the element in the first column as the left factor of the product) has the form

$$
\begin{array}{c|cccc}
 & 1 & b_1 & b_2 & b_1 \diamond b_2 \\
\hline
1 & 1 & b_1 & b_2 & b_1 \diamond b_2 \\
b_1 & b_1 & -1 & b_1 \diamond b_2 & -b_2 \\
b_2 & b_2 & -b_1 \diamond b_2 & -1 & b_1 \\
b_1 \diamond b_2 & b_1 \diamond b_2 & b_2 & -b_1 & -1 \ .
\end{array}
\tag{1.17}
$$

C_2 is referred to as the algebra of quaternions. One can easily verify that C_2^+ is isomorphic to C_1.

- C_3 has dimension 8 and the basis

$$\{1, b_1, b_2, b_3, b_1 \diamond b_2, b_2 \diamond b_3, b_1 \diamond b_3, b_1 \diamond b_2 \diamond b_3\} \ .$$

The full multiplication table can be constructed without difficulty. Let us take a closer look at the even subalgebra C_3^+. With its basis denoted by

$$\{1, b_2 \diamond b_3, b_3 \diamond b_1, b_1 \diamond b_2\} =: \{e_0, e_1, e_2, e_3\} \ , \tag{1.18}$$

the multiplication table is

$$
\begin{array}{c|cccc}
 & e_0 & e_1 & e_2 & e_3 \\
\hline
e_0 & e_0 & e_1 & e_2 & e_3 \\
e_1 & e_1 & -e_0 & e_3 & -e_2 \\
e_2 & e_2 & -e_3 & -e_0 & e_1 \\
e_3 & e_3 & e_2 & -e_1 & -e_0
\end{array}
\ .
\tag{1.19}
$$

Hence, the table is the same as (1.17), and the even subalgebra C_3^+ is isomorphic to the quaternion algebra C_2.

Now, let us consider an invertible element p of the Clifford algebra such that $p \diamond x \diamond p^{-1}$ is in X_1 for all x belonging to X_1. All such elements of the algebra constitute a multiplicative group. The transformation

$$O_p(x) := \alpha_p \, p \diamond x \diamond p^{-1} \ , \qquad \alpha_p = \pm 1 \tag{1.20}$$

is orthogonal $g(O_p(x), O_p(x)) = g(p \diamond x \diamond p^{-1}, p \diamond x \diamond p^{-1}) = -p \diamond x \diamond p^{-1} \diamond p \diamond x \diamond p^{-1} = p \diamond (-x \diamond x) \diamond p^{-1} = g(x, x)$. For a non–zero vector y of X_1 (i.e., invertible element of the algebra), the transformation O_y with $\alpha_y = -1$ is a reflection with respect to the hyperplane orthogonal to y: from (1.12) and (1.16) we have $g(y, y)(x \diamond y + y \diamond x) = 2g(x, y)y \diamond y$, and the right side Clifford product with y^{-1} gives $g(y, y)y \diamond x \diamond y^{-1} = -g(y, y)x + 2g(x, y)y$ or

$$-y \diamond x \diamond y^{-1} = x - \frac{2g(x, y)}{g(y, y)} y \ ;$$

cf. Eq.(1.5). Since orthogonal transformations are generated by reflections (Cartan theorem), it is natural to focus on Clifford products of the form $(-1)^m y_m \diamond \ldots \diamond y_2 \diamond y_1$, where y_k are non–zero vectors of X_1. Linearly dependent vectors y_k and λy_k correspond to the same reflection. Therefore, y_k are assumed to be unit vectors ($g(y_k, y_k) = 1$). The set containing the products of all sequences

$$y_m \diamond ... \diamond y_2 \diamond y_1 \ , \quad y_k \ \text{in} \ X_1 \ , \quad k = 1, ..., m$$

of unit vectors y_k with Clifford product constitutes a group denoted by $Pin(N)$. The group $Spin(N)$ is the subgroup of $Pin(N)$ restricted to $\mathcal{C}_N^+$, i.e., it consists of elements being products of even number of vectors. Since transformations expressible by even number of reflections are special orthogonal, the $Spin(N)$ group is related to $SO(N)$.

Two elements of $Pin(N)$ p and $-p$ represent the same orthogonal transformation, and there are no other elements representing that transformation. Therefore, there is a two–to–one homomorphism of $Pin(N)$ onto $O(N)$. Analogous statement can be made about $Spin(N)$ and $SO(N)$: there exists a two–to–one homomorphism of $Spin(N)$ onto $SO(N)$. This is usually expressed briefly by saying that $Pin(N)$ and $Spin(N)$ are two–fold coverings of $O(N)$ and $SO(N)$, respectively.

By definition of Pin, elements of the group $Pin(3)$ are of the form

$$(c^i b_i) \diamond (d^i b_i) \diamond ... \diamond (f^i b_i)$$

with $c^i c^i = 1$, $d^i d^i = 1$, ..., $f^i f^i = 1$, and with summation over $i = 1, 2, 3$. It can be verified by direct calculation that these products have two shapes:
– if the number of factors is odd, the element of $Pin(3)$ takes the form

$$a^i b_i + a^0 b_1 \diamond b_2 \diamond b_3 \overset{\bullet}{} ,$$

– if the number of factors is even, the element of $Pin(3)$ takes the form

$$a^0 + a^1 b_2 \diamond b_3 + a^2 b_3 \diamond b_1 + a^3 b_1 \diamond b_2 \ ; \tag{1.21}$$

in both cases the coefficients (a^0, a^1, a^2, a^3) satisfy $a^\mu a^\mu = 1$ (summation over $\mu = 0, 1, 2, 3$.)

The group $Spin(3)$ contains products of even number of factors. Based on (1.18) and (1.21) its elements can be written as

$$a^\mu e_\mu \ , \quad \text{with} \ a^\mu a^\mu = 1 \ .$$

We will return to this representation of special orthogonal transformations later.

2

Parameterizations

THE one–to–one correspondence between proper and improper rotations (based on the fact that in three dimensional space each improper rotation is a composition of the inversion and a unique proper rotation) allows us to limit the consideration of parameterizations to proper rotations. Using an additional discrete parameter pointing out whether the transformation involves the change of handedness or not, the improper rotations can be easily taken into account. Proper rotations can be parameterized by a 3×3 special orthogonal matrices. However, this parameterisation is highly redundant. As we mentioned earlier, because of the orthogonality conditions, the number of independent parameters is 3, and three parameters suffice to determine a rotation.

Is it possible to have a global "nice" (one–to–one, continuous, with continuous inverse) parameterization which would map the rotation group into the 3 dimensional Euclidean space? The answer is no, and this follows from topological arguments (Stuelpnagel, 1964).

A number of most frequently used parameterizations will be described in some detail. Many of them are constructed in such a way that singularities show up through multiple parameter sets corresponding to certain rotations. In most cases these rotations will be half–turns. As will be seen later, half–turns are in a sense special among rotations.

2.1 Half–turns

It is easy to see that if an orthogonal matrix O is symmetric, then the corresponding rotation is a half–turn. Apply O ($\neq I$) to both sides of the symmetry condition $O = O^T$. This gives $OO = I$. Thus, OO corresponds to the complete–turn and O corresponds to a half–turn.

The opposite statement is also true: if a rotation is a half–turn, it is represented by a *symmetric* special orthogonal matrix. Since a half–turn O applied twice gives the complete–turn which is equivalent to null rotation, the half–turn is identical with its inverse ($O = O^{-1}$). By its orthogonality, the matrix must be symmetric: $O = O^T$.

Moreover, the trace of a symmetric special orthogonal matrix is equal either to 3 or to -1. This can be easily demonstrated: Since the matrix is symmetric, its eigenvalues are real. Real eigenvalues of a special orthogonal matrix are equal to ± 1, at least one of them is equal to 1, and their product equals 1. Hence, if all eigenvalues are equal to 1, the trace is 3. If one eigenvalue equals 1 and the other two are -1, the trace is -1. Obviusly, only the identity matrix has the trace of 3. It will be evident later that the half–turns are the only proper rotations with the corresponding matrices having the trace of -1.

2.2 Cayley Transformation and Rodrigues Parameters

Let $x' = Ox$, where x is a vector and O is a special orthogonal matrix. The invariance of the vector's magnitude $x' \cdot x' = x \cdot x$ can be written as the orthogonality condition $y \cdot z = 0$ of vectors y and z defined by $y = x - x' = (I-O)x$ and $z = x+x' = (I+O)x$. For non–singular $I+O$, x can be expressed as $x = (I+O)^{-1}z$. Hence, the relation between y and z has the form $y = Rz$, where

$$R = (I - O)(I + O)^{-1} \ . \tag{2.1}$$

The orthogonality of y and z leads to $z \cdot Rz = 0$ and this occurs for all z. Hence, the matrix R is antisymmetic (skew–symmetric). Equation (2.1) can be transformed to $R+O+RO-I = 0$. Thus, for R satisfying $\det(I+R) \neq 0$,

$$O = (I + R)^{-1}(I - R) \ . \tag{2.2}$$

Alternatively, the relations (2.1) and (2.2) can be written in a different but fully equivalent way. If in $O = (O^{-1})^T$, the O on the right–hand side is substituted using (2.2), the matrix O is expressed as

$$O = (I - R)(I + R)^{-1} \ . \tag{2.3}$$

After transforming (2.3), we get

$$R = (I + O)^{-1}(I - O) \ . \tag{2.4}$$

Thus, for a special orthogonal matrix O such that $\det(I + O) \neq 0$, there exists a unique antisymmetric matrix R such that O is given by Eq.(2.3). This correspondence constitutes the so–called Cayley transformation (or factorization). For special orthogonal matrices the condition $\det(I + O) \neq 0$ limiting applicability of the Cayley transformation is equivalent to $\mathrm{tr}(O) \neq -1$; i.e., the transformation makes sense for matrices of all rotations except half–turns.

Equivalence of $\det(I+O) = 0$ and $1 + \mathrm{tr}(O) = 0$
The above statement follows from a more general formula. Let A and B be real nonsingular 3×3 matrices, and let C be defined by $C := A^{-1}B$. Then, it can be shown that

$$\det(\alpha A + \beta B) = \alpha^3 \det(A) + \beta^3 \det(B) + \alpha^2 \beta \det(A)\mathrm{tr}(C) + \alpha\beta^2 \det(B)\mathrm{tr}(C^{-1}) \ . \tag{2.5}$$

For special orthogonal matrices O_1 and O_2 and $O = O_1^{-1}O_2$, because of $\text{tr}(O^T) = \text{tr}(O)$ and $\alpha^3 + \beta^3 = (\alpha + \beta)(\alpha^2 - \alpha\beta + \beta^2)$, there occurs

$$\det(\alpha O_1 + \beta O_2) = (\alpha + \beta)(\alpha^2 + \beta^2 + \alpha\beta(\text{tr}(O) - 1)) .$$

In particular

$$\det(I + O) = 2(1 + \text{tr}(O)) \tag{2.6}$$

and it is different from 0 only if $\text{tr}(O) \neq -1$.

Moreover, one has $\det(O_1 - O_2) = 0$. In particular, $\det(O - I) = 0$. This provides another proof of the Euler theorem because it shows that the system of equations $Ox = x$ has a non–zero solution. $\boxtimes$

▫ *Auxiliary relations*
Further on, the following expression of the product of two permutation symbols will be needed:

$$\varepsilon_{ijk}\varepsilon_{lmn} = \begin{bmatrix} \delta_{il} & \delta_{jl} & \delta_{kl} \\ \delta_{im} & \delta_{jm} & \delta_{km} \\ \delta_{in} & \delta_{jn} & \delta_{kn} \end{bmatrix} . \tag{2.7}$$

Its particular forms

$$\varepsilon_{ijk}\varepsilon_{imn} = \delta_{jm}\delta_{kn} - \delta_{jn}\delta_{km} , \quad \varepsilon_{ijk}\varepsilon_{ijn} = 2\delta_{kn} , \quad \text{and} \quad \varepsilon_{ijk}\varepsilon_{ijk} = 6 \tag{2.8}$$

will be frequently used throughout the text.

From the definition of the matrix determinant, for a given matrix X,

$$\varepsilon_{sjk}X_{sn}X_{jl}X_{km} = \det(X)\varepsilon_{lmn} . \tag{2.9}$$

If X is non–singular and X^{-1} is its inverse, then

$$\varepsilon_{ijk}X_{jl}X_{km} = (\varepsilon_{sjk}X_{jl}X_{km})X_{sn}(X^{-1})_{ni} = \det(X)\varepsilon_{lmn}(X^{-1})_{ni} \tag{2.10}$$

Thus, the matrix X^{-1} can be expressed as $(X^{-1})_{ij} = \varepsilon_{ikl}\varepsilon_{jmn}X_{mk}X_{nl}/(2\det(X))$. In particular, we have

$$\varepsilon_{jst}O_{ji} = \varepsilon_{ikl}O_{sk}O_{tl} \tag{2.11}$$

for a 3×3 special orthogonal matrix O. $\boxtimes$

An $N \times N$ antisymmetric matrix R $(R^T + R = 0)$ is determined by $N(N-1)/2$ off–diagonal entries on one side of the diagonal. In the case of $N = 3$ the antisymmetric matrix R can be expressed as

$$R_{ij} = \varepsilon_{ijk}r^k \quad \text{with} \quad r^i = \frac{1}{2}\varepsilon_{ijk}R_{jk} , \tag{2.12}$$

where ε_{ijk} denotes the permutation symbol. Thus, rotations corresponding to matrices O such that $\text{tr}(O) \neq -1$ are parameterized by the three numbers r^i. They are usually called Rodrigues parameters. As a set, they are also referred to as 'Gibbs vector' (e.g., Korn & Korn, 1968).

The Rodrigues parameters can be expressed in a relatively simple way via the entries of O. It follows from (2.3) that

$$r^i = \frac{1}{2}\varepsilon_{ijk}R_{jk} = \frac{1}{2}\varepsilon_{ijk}((I+O)^{-1})_{jl}(I-O)_{lk} = \frac{1}{\det(I+O)}\varepsilon_{lmn}(\delta_{ni}+O_{ni})O_{ml} ;$$

equation (2.10) was used to eliminate $(I + O)^{-1}$. Based on (2.11) the term $\varepsilon_{lmn}O_{ni}O_{ml}$ is equal to $-\varepsilon_{iml}O_{ml}$. Thus, (2.6) leads to

$$r^i = -\frac{1}{1 + O_{ll}}\varepsilon_{ijk}O_{jk} \ . \tag{2.13}$$

The inverse relation can be obtained in a similar way

$$O_{ij} = (I - R)_{ik}((I + R)^{-1})_{kj} = \frac{1}{\det(I + R)}((1 - r^k r^k)\delta_{ij} + 2r^i r^j - 2\varepsilon_{ijk}r^k) \ .$$

Hence, the entries of the matrix O can be expressed as

$$O_{ij} = \frac{1}{1 + r^l r^l}((1 - r^k r^k)\delta_{ij} + 2r^i r^j - 2\varepsilon_{ijk}r^k) \ , \tag{2.14}$$

where (2.5) was used to replace $\det(I + R)$ by $1 + r^l r^l$.

In the Rodrigues parameterization, the identity rotation is given by $r^i = 0$, and the rotation inverse to r^i corresponds to $-r^i$. Both facts follow directly from (2.13). Moreover, it is easy to see that a half–turn would correspond to two mutually opposite Gibbs vectors of infinite magnitude.

Also the composition of rotations is expressible in a relatively simple way through Rodrigues parameters. Let us consider the composition $O''O'$ with O' and O'' corresponding to finite Rodrigues parameters r'^i and r''^i, respectively. If $r'^l r'''^l \neq 1$, then the parameters r^i corresponding to the product $O''O'$, are related to r'^i and r''^i via

$$r^i = (r'^i + r''^i - \varepsilon_{ijk}r'^j r''^k)/(1 - r'^l r'''^l) \ . \tag{2.15}$$

The derivation of the above formula is based on (2.13) and (2.14). For r corresponding to $O''O'$, one has $r^i = -\varepsilon_{ijk}O''_{jl}O'_{lk}/(1 + O''_{st}O'_{ts})$. After replacing components of matrices O' and O'' using (2.14) and simple but tedious calculations, the expression (2.15) is obtained. In more compact notation without subscripts, the operation (2.15) will be written as $r = r'' \circ r'$.

Eq.(2.15) in terms of antisymmetric matrices

Equation (2.15) can be used to justify the following statement: If O' and O'' are two 3×3 special orthogonal matrices given by Cayley transformations (2.3) $O' = (I - R')(I + R')^{-1}$ and $O'' = (I - R'')(I + R'')^{-1}$, and their product $O = O''O'$ is given by $O = (I - R)(I + R)^{-1}$, then the matrix R is related to R' and R'' by

$$R = (R' + R'' - [R', R''])/(1 + \mathrm{tr}(R'R'')/2) \ .$$

As in the case of Eq.(2.15), this relation is valid if none of the involved rotations is a half–turn. ⊠

Another way of operating on a vector

In relation to Eq.(2.12), it is worth to make the following note. Let x and x' be vectors be such that $x' = Ox$, where O represents a proper rotation. Then, the matrices $X_{ij} = \varepsilon_{ijk}x_k$ and $X'_{ij} = \varepsilon_{ijk}x'_k$ are linked via

$$X' = OXO^T \ . \tag{2.16}$$

This can be shown by using (2.11) and the orthogonality of O: $X'_{ij} = \varepsilon_{ijk}x'_k = \varepsilon_{ijk}O_{ks}x_s = \varepsilon_{mnl}O_{im}O_{jn}x_l = O_{im}O_{jn}X_{mn}$. Moreover, the matrix corresponding to the vector $z = -x \times y = Xy$ can be written as $Z = [X, Y] = XY - YX$, where $Y_{ij} = \varepsilon_{ijk}y_k$ and $Z_{ij} = \varepsilon_{ijk}z_k$. $\boxtimes$

Higher order Cayley transformations
As the power of an orthogonal matrix is an orthogonal matrix, for antisymmetric R and positive integer m, the matrix

$$((I - R)(I + R)^{-1})^m = (I - R)^m(I + R)^{-m} \qquad (2.17)$$

is special orthogonal. Tsiotras *et al.* (1997) call this mapping of antisymmetric matrices in the group of special orthogonal "the m–th order Cayley transformation". $\boxtimes$

Cayley transformation in other dimensions
When proper orthogonal transformations are considered in N dimensional space without limiting N to 3, a question about explicit parameterizations of such transformations arises. As Cayley transformation is valid for any dimension $N > 1$, the $N(N - 1)/2$ independent entries of R can be used as rotation parameters in spaces of dimension other than three. $\boxtimes$

2.3 Axis and Angle Parameters

Because of the Euler theorem, the parameterization by the rotation axis and the rotation angle (as the magnitude of rotation) is very natural. However, within this approach, there are no unique rotation axes for half–turns and null rotations.

The *rotation angle* is expressible via entries of the orthogonal matrix O corresponding to the rotation. Let the unit vector in the direction of the rotation axis be denoted by n. The vectors $v^{(i)}$ defined by $v_j^{(i)} := \varepsilon_{ijk}n_k$ are perpendicular to n because $v_j^{(i)}n_j = \varepsilon_{ijk}n_jn_k = 0$. The rotation angle ω corresponding to O is equal to the angle between $v^{(i)}$ and its image $Ov^{(i)}$. Thus, the scalar product of $v^{(i)}$ and $Ov^{(i)}$ is equal to $|v^{(i)}|^2 \cos\omega$. Therefore, we can write

$$v_j^{(i)}v_j^{(i)} \cos\omega = O_{mn}v_m^{(i)}v_n^{(i)} \ .$$

Based on the definition of $v^{(i)}$, this can be transformed to $\varepsilon_{ijs}n_s\varepsilon_{ijt}n_t \cos\omega = O_{mn}\varepsilon_{ims}n_s\varepsilon_{int}n_t$. Hence, after using (2.8), one obtains

$$2\cos\omega = O_{mm} - 1 \ . \qquad (2.18)$$

This relates the cosine of the rotation angle to the trace of the rotation matrix.

Is the angle ω fully determined by Eq.(2.18)? The relation involves cosine of ω, so the answer is 'yes' only if ω is in the range $[0, \pi]$. Is it possible to keep ω in that range and cover all rotations? The answer is positive if n is allowed to take any direction in space. It is also noteworthy that, with ω in $[0, \pi]$, the traces of 3 and -1 correspond to $\omega = 0$ and $\omega = \pi$, respectively.

Our next step is to link the matrix O with the *rotation axis*. At the outset, let us limit our considerations to rotations represented by special orthogonal matrices with traces different from -1 and 3. In such a case, the Rodrigues parameters are non–zero and finite. The Gibbs vector is an eigenvector of the corresponding rotation matrix.

Evidence
It follows from (2.13) that

$$O_{ji}r^j = -\frac{1}{1+O_{ll}}\varepsilon_{jmn}O_{ji}O_{mn} .$$

Using (2.11) and again (2.13), one gets $O_{ji}r^j = r^i$. By applying O_{ki} to both sides of the last relation, the desired result $O_{ki}r^i = r^k$ is obtained. ⊠

It is a standard to represent the rotation axis by a unit vector $n = r/\sqrt{r^k r^k}$, which can be done for $\sqrt{r^k r^k} \neq 0$. From (2.13), we have $r^k r^k = (3 - O_{jk}O_{kj}/(1+O_{ll})^2$. Thus,

$$n_i = \frac{-1}{\sqrt{3 - O_{lm}O_{ml}}}\varepsilon_{ijk}O_{jk} . \qquad (2.19)$$

The expression is valid for O such that $O_{lm}O_{ml} \neq 3$ which is equivalent to the initial condition that $\text{tr}(O)$ differs from -1 and 3.

To express the rotation matrix through the considered parameters, both n and ω must be involved. Let us notice that

$$\sqrt{r^l r^l} = \frac{\sqrt{3 - O_{jk}O_{kj}}}{1+O_{ll}} = \frac{\sqrt{3 - (2\cos(2\omega) + 1)}}{1 + (2\cos\omega + 1)} = \frac{|\sin\omega|}{1 + \cos(\omega)} .$$

For ω in $[0, \pi)$, the right hand side equals $\tan(\omega/2)$, and the Gibbs vector can be expressed as

$$r = n\tan(\omega/2) . \qquad (2.20)$$

With r^l in (2.14) substituted by $n_l \tan(\omega/2)$, the entries of O are given by the (Rodrigues) formula

$$O_{ij} = \delta_{ij}\cos\omega + n_i n_j(1 - \cos\omega) - \sin\omega\varepsilon_{ijk}n_k . \qquad (2.21)$$

Although, the relations above were obtained for rotations by angles different than 0 or π, the formula (2.21) is general, i.e., it is also valid for these particular values of ω. Consideration of the limiting cases confirms that the rotation by $\omega = 0$ is represented by the identity matrix, and shows that the rotations by π correspond to

$$O_{ij} = O_{ji} = 2n_i n_j - \delta_{ij} . \qquad (2.22)$$

The transformed version of this relation $n_i n_j = (\delta_{ij} + O_{ij})/2$ serves as the way of determining the axis direction from O in the special case of $\omega = \pi$; the solution is not unique as the rotations by π about n and $-n$ give the same result. Moreover, it is easy to notice that (2.21) provides a correct orthogonal matrix even if ω is not limited to $[0, \pi]$.

Finally, it is important to be aware of the relationship between rotations by the same angle about different axes. Let the rotation about n by the angle ω be denoted by $O(n, \omega)$. Using the Rodrigues formula (2.21), it can be shown that the rotation $O'O(n, \omega)O'^{-1}$ conjugate to $O(n, \omega)$ is equivalent to the rotation about the axis $O'n$ by the same angle ω, i.e.,

$$O'O(n, \omega)O'^{-1} = O(O'n, \omega) . \tag{2.23}$$

The proof of this statement is not difficult.

□ *Axes of conjugate rotations*
Due to (2.21),

$$O'_{mi}O_{ij}(n, \omega)O'_{nj} = \delta_{ij} \cos \omega + O'_{mi}O'_{nj}n_in_j(1 - \cos \omega) - \sin \omega \varepsilon_{ijk}O'_{mi}O'_{nj}n_k .$$

Using (2.11), the last term is transformed in such a way that

$$O'_{mi}O_{ij}(n, \omega)O'_{nj}$$
$$= \delta_{ij} \cos \omega + (O'_{mi}n_i)(O'_{nj}n_j)(1 - \cos \omega) - \sin \omega \varepsilon_{mnl}(O'_{lk}n_k) = O_{mn}(O'n, \omega) .$$

This proves the statement about conjugate rotations.
     The following is worth mentioning: For the rotation by π about m and the rotation by an arbitrary angle ω about n perpendicular to m, there occurs

$$O(m, \pi)O(n, \omega)O(m, \pi) = O(m, \pi)O(n, \omega)O(-m, \pi) = O(-n, \omega) . \tag{2.24}$$

To explain the last step it is enough to use the formula for conjugate rotations (2.23) and to note that the rotation $O(m, \pi)$ reverses the direction of n, i.e., $O(m, \pi)n = -n$. ⊠

□ *Half–turn relating two vectors*
     This is probably the best place to make the following digression: Let us take two unit vectors x and y such that $x \neq -y$. The vector x can be transformed to the position of y using the matrix O defined by

$$O_{ij} = \frac{1}{1 + x_ly_l}(x_i + y_i)(x_j + y_j) - \delta_{ij} . \tag{2.25}$$

Simple calculation shows that $O_{ij}x_j = y_i$ so the above statement is correct. It is also straightforward to check that O is special orthogonal, symmetric, and that $\operatorname{tr}(O) = -1$; i.e., it represents a half–turn. Obviously, such O is not the only special orthogonal matrix having that property. However, for $x \neq \pm y$ it is a unique *symmetric* special orthogonal matrix transforming x into y. Because of symmetry, it also transforms y into x. ⊠

□ *Representation of improper rotations*
     In relation to Eq.(2.22), let us notice that the improper transformation of reflection with respect to a plane perpendicular to a unit vector n is represented by the orthogonal matrix

$$\delta_{ij} - 2n_in_j . \tag{2.26}$$

This is easily understandable because the reflection can be seen as the composition of the inversion and the rotation about n by π.
     A general improper rotation is a composition of a proper rotation and the reflection with respect to the plane perpendicular to the rotation axis. If the rotation

is represented by the matrix O and the rotation axis is represented by n, then the orthogonal matrix P corresponding to the improper rotation has the form

$$P_{ij} = (\delta_{ik} - 2n_i n_k)O_{kj} = O_{ij} - 2n_i n_k \ .$$

On the other hand, the improper rotation can be also expressed as a composition of a proper rotation, say O', and the inversion $\mathfrak{I}$: $P = \mathfrak{I}O'$. The matrices O and O' are related by

$$O'_{ij} = 2n_i n_k - O_{ij} \ ;$$

this leads to $n' = -n$ and $\omega' = \pi - \omega$, where n and ω are the rotation axis and angle corresponding to O, and n' and ω' correspond to O'. $\boxtimes$

Reflections and Cayley factorization

As was already mentioned, a proper rotation can be decomposed into two reflections. Since reflections are represented by matrices of the type (2.26), a special orthogonal matrix O can be written in the form

$$O = (I - 2L)(I - 2K) = (I - 2L)(I - 2K)^{-1} \ , \qquad (2.27)$$

where $K_{ij} = k_i k_j$ and $L_{ij} = l_i l_j$, and k and l are two unit vectors perpendicular to the reflection planes. The question is what is the relationship of the above decomposition to the Cayley factorization. Simple calculation shows that for nonsingular $I - (L + K)$, the matrix

$$R = (L - K)(I - (L + K))^{-1}$$

is antisymmetric, and it is related to O via Cayley transformation (2.3). As the matrix R in the Cayley factorization is unique and the decomposition into reflections is not, there are multiple pairs of K, L matrices leading to the same R. $\boxtimes$

Composition of rotations via decomposition into reflections

It follows from Eqs. (2.27) and (2.18) that $\cos(\omega/2) = |k_i l_i|$, i.e., the angle between the two reflecting planes equals $1/2$ of the rotation angle. The null rotation corresponds to $K = L$. For non–trivial rotations, the rotation axis (determined from Eqs. (2.19) or (2.22)) is perpendicular to the vectors k and l. A rotation given by axis and angle is decomposed into two reflections by selecting an arbitrary plane containing the rotation axis as one of the two mirrors; the other plane of reflection is determined by the axis and the angle to the first plane.

Let us consider a composition of two rotations with axes and angles (n, ω) and (n', ω'). The next step is to decompose each of the rotations into two reflections $O(n, \omega) = (I - 2L)(I - 2K)$ and $O(n', \omega') = (I - 2L')(I - 2K')$. (The sense of these relations being the same as that of Eq.(2.27).) Now, the arbitrariness in the choice of the reflecting planes allows the second reflection plane in decomposition of (n, ω) to be the same as the first mirror of (n', ω'). Geometrically, this means that the plane contains both n and n'. The choice can be expressed as $K' = L$. In consequence,

$$O(n', \omega')O(n, \omega) = (I - 2L')(I - 2K')(I - 2L)(I - 2K) = (I - 2L')(I - 2K) \ ,$$

i.e., the resultant rotation $O(n', \omega')O(n, \omega)$ is equivalent to the composition of reflections with respect to the planes corresponding to K and L'. Hence, the axis of the resultant rotation is the intersection of the planes corresponding to K and L' with the sense determined by the right hand rule (the smaller turn of K plane towards L' plane). The angle of the resultant rotation is twice the angle between the planes corresponding to K and L'. See (Fig. 2.1). The above line of reasoning is another way of justifying the Rodrigues–Hamilton theorem. $\boxtimes$

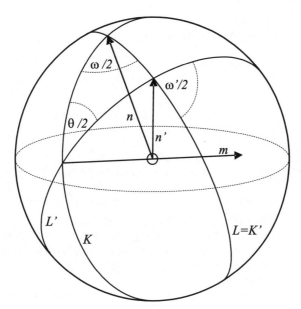

Fig. 2.1. Ilustration of the composition of rotations via reflections. Great circles on the sphere are traces of the reflection planes. They are marked by the same symbols as the matrices associated with the normals to the planes. The resultant rotation is about m by θ.

2.4 Euler Angles

In texture analysis, the so–called Euler angles are by far the most frequently used rotation parameters. The reasons are mostly historical; they are convenient for series expansion of functions defined on rotations, and that approach was considered to be important at the early stages of the quantitative texture analysis.

Let $O(n,\omega)$ be an orthogonal matrix corresponding to the rotation about n by the angle ω. Moreover, let e_i be vectors determining the axes of the Cartesian coordinate system fixed in space, i.e., $e_1 = [1\,0\,0]^T$, $e_2 = [0\,1\,0]^T$ and $e_1 = [0\,0\,1]^T$. Then, according to (2.21)

$$O(e_1,\omega) = \begin{bmatrix} 1 & 0 & 0 \\ 0 & \cos\omega & \sin\omega \\ 0 & -\sin\omega & \cos\omega \end{bmatrix}, \quad O(e_2,\omega) = \begin{bmatrix} \cos\omega & 0 & -\sin\omega \\ 0 & 1 & 0 \\ \sin\omega & 0 & \cos\omega \end{bmatrix}$$

$$\text{and} \quad O(e_3,\omega) = \begin{bmatrix} \cos\omega & \sin\omega & 0 \\ -\sin\omega & \cos\omega & 0 \\ 0 & 0 & 1 \end{bmatrix}.$$

The matrix obtained as the composition of the matrices $O(e_3,\varphi_1)$, $O(e_1,\phi)$ and $O(e_3,\varphi_2)$ will be denoted as $O(\varphi_1,\phi,\varphi_2)$; its explicit form is

$$O(\varphi_1,\phi,\varphi_2) = O(e_3,\varphi_2)O(e_1,\phi)O(e_3,\varphi_1)$$

$$
= \begin{bmatrix} \cos\varphi_1 \ \cos\varphi_2 - \sin\varphi_1 \ \sin\varphi_2 \ \cos\phi \\ -\cos\varphi_1 \ \sin\varphi_2 - \sin\varphi_1 \ \cos\varphi_2 \ \cos\phi \\ \sin\varphi_1 \ \sin\phi \end{bmatrix}
$$

$$
\begin{bmatrix} \sin\varphi_1 \ \cos\varphi_2 + \cos\varphi_1 \ \sin\varphi_2 \ \cos\phi & \sin\varphi_2 \ \sin\phi \\ -\sin\varphi_1 \ \sin\varphi_2 + \cos\varphi_1 \ \cos\varphi_2 \ \cos\phi & \cos\varphi_2 \ \sin\phi \\ -\cos\varphi_1 \ \sin\phi & \cos\phi \end{bmatrix}.
$$

To each set of angles $(\varphi_1, \phi, \varphi_2)$ corresponds an orthogonal matrix. Only trigonometric functions of the angles are involved in this relationship, therefore, the range of the angles can be limited to $0 \leq \varphi_1, \phi, \varphi_2 < 2\pi$. However, brief calculation based on (2.24) shows that $O(\varphi_1, \phi, \varphi_2) = O(\pi + \varphi_1, 2\pi - \phi, \pi + \varphi_2)$. Thus, the range given above must be limited; the convention is to take

$$
0 \leq \varphi_1 < 2\pi \ , \quad 0 \leq \phi \leq \pi \ , \quad 0 \leq \varphi_2 < 2\pi \ . \tag{2.28}
$$

For $\varphi_1, \phi, \varphi_2$ to be convenient parameters, the above ranges must cover all possible proper rotations. The coverage occurs if the angles within (2.28) are expressible via entries of an arbitrary orthogonal matrix. It is easy to see that the latter is true. If the entry O_{33} is neither 1 nor -1, then

$$
\begin{aligned}
\sin\varphi_1 &= O_{31}/\zeta \ , \quad \cos\varphi_1 = -O_{32}/\zeta \ , \\
\cos\phi &= O_{33} \ , \\
\sin\varphi_2 &= O_{13}/\zeta \ , \quad \cos\varphi_2 = O_{23}/\zeta \ ,
\end{aligned}
$$

where $\zeta = \sqrt{1 - (O_{33})^2}$. Due to the orthogonality of O, the entry O_{33} is in the range $[-1, 1]$, and the quantities $(O_{31}/\zeta)^2 + (-O_{32}/\zeta)^2$ and $(O_{13}/\zeta)^2 + (O_{23}/\zeta)^2$ are equal to 1. Thus, the above formulas are consistent, and correctly determine the angles φ_1, ϕ and φ_2. If $O_{33} = \cos\phi = 1$, then

$$
\sin(\varphi_1 + \varphi_2) = O_{12} \ , \quad \cos(\varphi_1 + \varphi_2) = O_{11} \ .
$$

Similarly, if $O_{33} = \cos\phi = -1$, then

$$
\sin(\varphi_1 - \varphi_2) = O_{12} \ , \quad \cos(\varphi_1 - \varphi_2) = O_{11} \ .
$$

This shows that the correspondence between the matrices and Euler angles is unambiguous only if $\phi \neq 0$ and $\phi \neq \pi$. If $\phi = 0$, only the sum of φ_1 and φ_2 is uniquely related to the rotation. One can write $O(\varphi_1, 0, \varphi_2) = O(\varphi_1 + \alpha, 0, \varphi_2 - \alpha)$, where α is arbitrary, and addition and subtraction of α are modulo 2π. Analogously, if $\phi = \pi$, the difference between φ_1 and φ_2 is uniquely related to the rotation. In that case, $O(\varphi_1, \pi, \varphi_2) = O(\varphi_1 + \alpha, \pi, \varphi_2 + \alpha)$. In engineering literature, this ambiguity of Euler angles is referred to as "gimbal lock".

It is trivial to check that $O(0, 0, 0)$ represents identity. The matrix inverse to $O(\varphi_1, \phi, \varphi_2)$ is given by $O(-\varphi_2, -\phi, -\varphi_1)$; for the angles in the proper range (2.28), the inverse is equal to $O(\xi(\varphi_2), \phi, \xi(\varphi_1))$, with $\xi(x) = \pi - x$ for $x \leq \pi$ and $\xi(x) = 3\pi - x$ for $x > \pi$.

A related parameterization based on the Euler angles and occasionally used in texture analysis was originally proposed by Lattman (1972). The parameters are $(\varphi^+, \phi, \varphi^-)$, where

$$
\varphi^+ = (\varphi_1 + \varphi_2)/2 \quad \text{and} \quad \varphi^- = (\varphi_1 - \varphi_2)/2 \ .
$$

Obviously, this modification does not remove singularities: At $\phi = 0$, orientation is specified by φ^+ and the value of φ^- is immaterial. Similarly, at $\phi = \pi$, φ^- determines the orientation and the value of φ^+ does not play any role. These new parameters were promoted because they simplify the interpretation of texture components in graphs of orientation distributions. However, their use remains limited.

Another way of introducing Euler angles
    The Euler angles are sometimes introduced in a geometric way by composing rotations about axes of the coordinate system attached to the object, i.e., the system which is displaced by the rotation. Let e_1' be the image of e_1 in the rotation $O(e_3, \varphi_2)$, and let e_3'' be the image of e_3 in the rotation $O(e_1', \phi)O(e_3, \varphi_2)$. It can be shown that the rotation by φ_2 about e_3 followed by the rotation by ϕ about e_1' (the new position of e_1), followed the rotation by φ_1 about e_3'' (the newest position of e_3) is equivalent to $O(\varphi_1, \phi, \varphi_2)$, i.e.,

$$O(e_3, \varphi_2)O(e_1, \phi)O(e_3, \varphi_1) = O(e_3'', \varphi_1)O(e_1', \phi)O(e_3, \varphi_2) . \tag{2.29}$$

The proof is simple. Due to the relation (2.23) between conjugate elements of the rotation group,

$$O(e_1', \phi) = O(O(e_3, \varphi_2)e_1, \phi) = O(e_3, \varphi_2)O(e_1, \phi)O(-e_3, \varphi_2)$$

and

$$\begin{aligned} O(e_3'', \varphi_1) &= O(O(e_1', \phi)O(e_3, \varphi_2)e_3, \varphi_1) \\ &= O(e_1', \phi)O(e_3, \varphi_2)O(e_3, \varphi_1)O(-e_3, \varphi_2)O(-e_1', \phi) . \end{aligned}$$

It is enough to substitute $O(e_1', \phi)$ and $O(e_3'', \varphi_1)$ in the right–hand side of Eq.(2.29) to see its correctness. ☒

2.4.1 Other conventions

In the introduction of Euler angles a number of arbitrary choices were made. One of them was the choice of the axes e_3, e_1 and e_3. It follows certain convention which is used in texture analysis but is different from the convention currently most common in physics literature. The one used in physics differs by the second axis of rotation; it is e_2 (y–convention) instead of e_1 (x–convention). This influences the formulas but there are no essential differences in their derivation. Customarily, the angles of the y–convention are named α, β and γ, and their relationship to φ_1, ϕ and φ_2 is: $\varphi_1 = \alpha + \pi/2$, $\phi = \beta$ and $\varphi_2 = \gamma + 3\pi/2$ (e.g., Matthies, Vinel & Helming, 1987).

    Euler angles are sometimes defined in a considerably different way with the rotation axes e_1, e_2 and e_3. By definition, the orthogonal matrix associated with the parameters ϕ_1, ϕ_2 and ϕ_3 is

$$O(e_3, \phi_3)O(e_2, \phi_2)O(e_1, \phi_1) .$$

This expression looks symmetric but the advantage is fictitious: the symmetry in not transferred any further because of the non–comutative nature of the rotation group. (These parameters are sometimes referred to as Cardan angles or Bryant angles or Tait–Bryant angles. They correspond to "roll–pitch–yaw" angles used in aerodynamics.) One can easily verify that

$$O(e_3, \phi_3)O(e_2, \phi_2)O(e_1, \phi_1)$$

$$= \begin{bmatrix} \cos\phi_2 \cos\phi_3 & \cos\phi_1 \sin\phi_3 + \sin\phi_1 \sin\phi_2 \cos\phi_3 \\ -\cos\phi_2 \sin\phi_3 & \cos\phi_1 \cos\phi_3 - \sin\phi_1 \sin\phi_2 \sin\phi_3 \\ \sin\phi_2 & -\sin\phi_1 \cos\phi_2 \end{bmatrix}$$

$$\begin{bmatrix} \sin\phi_1 \sin\phi_3 - \cos\phi_1 \sin\phi_2 \cos\phi_3 \\ \sin\phi_1 \cos\phi_3 + \cos\phi_1 \sin\phi_2 \sin\phi_3 \\ \cos\phi_1 \cos\phi_2 \end{bmatrix}.$$

The parameterization covers the whole set of rotations. However, with the full range $[0, 2\pi)$ of angles, two different sets of parameters give the same matrix because $O(e_3, \phi_3)O(e_2, \phi_2)O(e_1, \phi_1) = O(e_3, \phi_3+\pi)O(e_2, \pi-\phi_2)O(e_1, \phi_1+\pi)$. If the ranges

$$0 \le \phi_1 < 2\pi , \quad -\pi/2 \le \phi_2 \le \pi/2 , \quad 0 \le \phi_3 < 2\pi$$

are assumed, the correspondence is unambiguous except the cases $\phi_2 = -\pi/2$ and $\phi_2 = \pi/2$ in which only $\phi_1 - \phi_3$ and $\phi_1 + \phi_3$ can be determined.

Other combinations of rotation axes
A number of other combinations of rotation axes are occasionally used in definitions of Euler angles (Pio, 1966). One of the uncommon conventions is claimed to have some advantages iIn crystallographic applications. The angles are customarily named φ, ϑ and ρ and the sequence is: i. rotation by φ about e_3, ii. rotation by ϑ about new e_1, and iii. rotation by ρ about the newest e_2. See, e.g., Scheringer, 1963 or Sussman *et al.*, 1977. $\boxtimes$

Generalized Euler angles
Parameters constructed analogously to Euler angles can be used to parametrize rotations in spaces of higher dimension. This can be done using a procedure described in the book by Murnaghan (1962). Another one was given by Hoffman, Raffenetti & Ruedenberg, (1972). $\boxtimes$

2.5 Cayley–Klein Parameters

We have not encountered any direct applications of the issues presented in this section to textures. However, the account on them is given here not only for completeness but also because of their close relationship to quaternions, which are a very useful tool for dealing with rotations. The interpretation of Cayley–Klein parameters comes from the relation between the rotations and some transformations of the Gauss' plane of complex numbers called linear fractional transformations.

2.5.1 Linear fractional transformations

In general, a linear fractional transformation (or homography or Möbius transformation) is a transformation of the complex plane of the form

$$t \to t' = \frac{at + b}{ct + d} , \quad ad - bc \ne 0 , \tag{2.30}$$

where a, b, c, d and t and t' are complex. Due to $ad - bc \neq 0$, the transformation is invertible. Without loosing generality, the condition $ad - bc \neq 0$ is replaced by $ad - bc = 1$. There is still some arbitrariness in the choice of the coefficients: the substitution $a, b, c, d \rightarrow -a, -b, -c, -d$ gives the same homography, and that is the only set of different parameters having this property.

The composition of two homographies is a homography. The composition of homographies can be represented by the multiplication of 2×2 matrices of unit determinant. Precisely, if the matrix $\begin{bmatrix} a & b \\ c & d \end{bmatrix}$ is assigned to the homography (2.30), and $\begin{bmatrix} a' & b' \\ c' & d' \end{bmatrix}$, $(a'd' - b'c' = 1)$ corresponds to $t' \rightarrow t'' = (a't' + b')/(c't' + d')$ then the appropriate matrix for $t \rightarrow t''$ is given by the product $\begin{bmatrix} a' & b' \\ c' & d' \end{bmatrix} \begin{bmatrix} a & b \\ c & d \end{bmatrix}$.

Antihomographies

In analogy, the so–called antihomographies can be considered. An antihomography is a transformation of the complex plane defined by

$$t \rightarrow t' = \frac{a\bar{t} + b}{c\bar{t} + d} , \quad ad - bc \neq 0 .$$

The composition of two antihomographies is a homography. The composition of homography and antihomography (in arbitrary order) is an antihomography. ⊠

2.5.2 Stereographic projection of sphere

A complex variable

$$t = x_1 + ix_2 = re^{i\beta} , \quad \text{where } x_1, x_2, r, \beta \quad \text{are real,}$$

can be mapped on the unit sphere

$$x_1^2 + x_2^2 + x_3^2 = 1$$

using the inverse stereographic projection: the image of a given point of the complex plane $x_3 = 0$ is given by the point of intersection of the sphere with the line through the given point and the south pole of the sphere; see Fig. 2.2. By assumption: the x_3 axis goes through north ($x_1 = 0 = x_2$, $x_3 = +1$, $t = 0$) and south ($x_1 = 0 = x_2$, $x_3 = -1$) poles of the sphere, $x_3 = 0$ on the equator, and real t are mapped on the meridian $x_2 = 0$.

Let the sphere be parameterized by longitude β and $\alpha = \pi/2 - $ latitude. The parameter r is related to α via $r = \tan(\alpha/2)$. The relations $x_1 = \cos\beta\sin\alpha$, $x_2 = \sin\beta\sin\alpha$ and $x_3 = \cos\alpha$ can be rewritten using the parameters r and β as

$$x_1 = \frac{2r\cos\beta}{1 + r^2} , \quad x_2 = \frac{2r\sin\beta}{1 + r^2} , \quad x_3 = \frac{1 - r^2}{1 + r^2} .$$

Since $r^2 = t\bar{t}$, $2r\cos\beta = t + \bar{t}$ and $2ir\sin\beta = t - \bar{t}$, we have

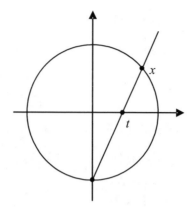

Fig. 2.2. Stereographic projection.

$$x_1 = \frac{t + \bar{t}}{1 + t\bar{t}}, \quad x_2 = -i\frac{t - \bar{t}}{1 + t\bar{t}}, \quad x_3 = \frac{1 - t\bar{t}}{1 + t\bar{t}}. \tag{2.31}$$

As for the inverse mapping, the complex parameter t can be expressed through x_1, x_2 and x_3

$$t = \frac{x_1 + ix_2}{1 + x_3} = \frac{1 - x_3}{x_1 - ix_2}. \tag{2.32}$$

The above relations are valid for all points of the sphere except the south pole.

2.5.3 Rotations and homographies

Based on the correspondence between the Gauss plane and the sphere, let us find out what will be the point t' obtained from t due to the rotation of the sphere by φ with the rotation axis being e_3. From (1.3) follows

$$\begin{bmatrix} x_1' \\ x_2' \\ x_3' \end{bmatrix} = \begin{bmatrix} \cos\varphi & \sin\varphi & 0 \\ -\sin\varphi & \cos\varphi & 0 \\ 0 & 0 & 1 \end{bmatrix} \begin{bmatrix} x_1 \\ x_2 \\ x_3 \end{bmatrix}.$$

From Eq.(2.32), the corresponding transformation is

$$t' = \frac{x_1' + ix_2'}{1 + x_3'} = \frac{x_1 + ix_2}{1 + x_3}(\cos\varphi - i\sin\varphi) = t\exp(-i\varphi) = \frac{t\exp(-i\varphi/2)}{\exp(i\varphi/2)},$$

i.e., a homography with the matrix $\begin{bmatrix} \exp(-i\varphi/2) & 0 \\ 0 & \exp(i\varphi/2) \end{bmatrix}$. Similarly, the rotation by ϕ about the e_1–axis corresponds to the homography

$$t' = \frac{x_1' + ix_2'}{1 + x_3'} = \frac{x_1 + ix_2\cos\phi + ix_3\sin\phi}{1 - x_2\sin\phi + x_3\cos\phi} = \frac{(t\cos(\phi/2) + i\sin(\phi/2))}{(it\sin(\phi/2) + \cos(\phi/2))}.$$

As we saw in the section on Euler angles, each rotation can be composed of rotations (e_3, φ_1), (e_1, ϕ) and (e_3, φ_2). On the other hand, the composition of homographies is a homography. Thus, for an arbitrary rotation there is a corresponding homography given by the matrix of the form

$$
\begin{bmatrix} \exp(-i\varphi_2/2) & 0 \\ 0 & \exp(i\varphi_2/2) \end{bmatrix} \begin{bmatrix} \cos(\phi/2) & i\sin(\phi/2) \\ i\sin(\phi/2) & \cos(\phi/2) \end{bmatrix} \begin{bmatrix} \exp(-i\varphi_1/2) & 0 \\ 0 & \exp(i\varphi_1/2) \end{bmatrix}
$$

$$
= \begin{bmatrix} \cos(\phi/2)\exp(-i(\varphi_1+\varphi_2)/2) & i\sin(\phi/2)\exp(i(\varphi_1-\varphi_2)/2) \\ i\sin(\phi/2)\exp(i(-\varphi_1+\varphi_2)/2) & \cos(\phi/2)\exp(i(\varphi_1+\varphi_2)/2) \end{bmatrix} . \tag{2.33}
$$

Is every homography on the Gauss plane a rotation on the sphere? The answer is 'no'. The unimodular matrix corresponding to the homography above is unitary and, therefore, all matrices of homographies representing rotations belong to $SU(2)$. Consequently, the homography corresponding to a rotation must satisfy the conditions (1.11)

$$
d = \bar{a} \quad \text{and} \quad c = -\bar{b} .
$$

The question asked before, can be reformulated: does every homography with a special unitary matrix correspond to a rotation on the sphere? This time the answer is positive.

Let us take such a homography with unimodular unitary matrix, i.e., with coefficients satisfying $d = \bar{a}$, $c = -\bar{b}$ and $a\bar{a} + b\bar{b} = 1$. Now,
– by expressing the number t by vector components x_1, x_2, x_3 using Eq.(2.32), then
– by applying the above homography to t to obtain t' ($t' = (at+b)/(-\bar{b}t+\bar{a})$), and finally,
– by calculating the components x_1', x_2', x_3' corresponding to t' via the analogue of (2.31), we get $x' = Ox$, where

$$
O = \begin{bmatrix} \frac{1}{2}(a^2+\bar{a}^2-b^2-\bar{b}^2) & \frac{i}{2}(a^2-\bar{a}^2+b^2-\bar{b}^2) & (ab+\bar{a}\,\bar{b}) \\ \frac{i}{2}(-a^2+\bar{a}^2+b^2-\bar{b}^2) & \frac{1}{2}(a^2+\bar{a}^2+b^2+\bar{b}^2) & -i(ab-\bar{a}\,\bar{b}) \\ -(a\bar{b}+\bar{a}b) & -i(a\bar{b}-\bar{a}b) & (a\bar{a}-b\bar{b}) \end{bmatrix} . \tag{2.34}
$$

Direct calculation confirms that the matrix O is special orthogonal and thus, the transformation is a proper rotation. The interrelated complex numbers a, b, c and d are referred to as Cayley–Klein parameters of a rotation.

Since a and b satisfy the condition $a\bar{a} + b\bar{b} = 1$, only three independent real numbers are involved. Let q^μ, $(\mu = 0, 1, 2, 3)$[1] be real numbers related to a and b through $a = q^0 + iq^3$ and $b = q^2 - iq^1$. The condition $a\bar{a} + b\bar{b} = 1$ takes the form

$$
q^\mu q^\mu = 1 . \tag{2.35}
$$

The numbers q^μ determine a unitary matrix and the corresponding rotation.

Now, let the unimodular unitary matrix U be given by

$$
U = \begin{bmatrix} q^0 + iq^3 & q^2 - iq^1 \\ -q^2 - iq^1 & q^0 - iq^3 \end{bmatrix} , \tag{2.36}
$$

---

[1] Greek indices are assumed to take values $0, 1, 2, 3$ and Latin indices are $1, 2, 3$.

and let the unimodular unitary matrices U' and $U'' = U'U$ be determined by q'^μ and q''^μ in analogous way. The relation $U'' = U'U$, or in extenso

$$\begin{bmatrix} q''^0 + iq''^3 & q''^2 - iq''^1 \\ -q''^2 - iq''^1 & q''^0 - iq''^3 \end{bmatrix} = \begin{bmatrix} q'^0 + iq'^3 & q'^2 - iq'^1 \\ -q'^2 - iq'^1 & q'^0 - iq'^3 \end{bmatrix} \begin{bmatrix} q^0 + iq^3 & q^2 - iq^1 \\ -q^2 - iq^1 & q^0 - iq^3 \end{bmatrix},$$
(2.37)

leads to

$$q''^0 = q^0 q'^0 - q^i q'^i , \qquad q''^i = q^0 q'^i + q'^0 q^i + \varepsilon_{ijk} q'^j q^k .$$
(2.38)

This is the explicit formula for composing rotations given by Cayley–Klein parameters a, b and a', b', where $a = q^0 + iq^3$, $b = q^2 - iq^1$ and $a' = q'^0 + iq'^3$, $b' = q'^2 - iq'^1$.

As was mentioned before, the change of sign

$$a \to -a , \quad b \to -b ,$$

does not affect the corresponding homography and rotation. However, it gives a new unitary matrix and a new set of parameters q^μ. Thus, the correspondence between these parameters and rotations is two–to–one. (The equivalence of the antipodal points q^μ and $-q^\mu$ of the sphere (2.35) tells a mathematician that the space of rotations has the topology of the three dimensional projective space).

Improper rotations and antihomographies
First of all, let us notice that it is impossible to express inversion $(x_1, x_2, x_3) \to (x'_1, x'_2, x'_3) = (-x_1, -x_2, -x_3)$ or any other improper rotation by a homography; the appropriate calculation gives mutually contradicting equations for the coefficients of the homography. This is not the case if the transformation is chosen to be an antihomography. To find the antihomography corresponding to the inversion, let us use the following sequence

$$t' = \frac{x'_1 + ix'_2}{1 + x'_3} = \frac{1 - x'_3}{x'_1 - ix'_2} = \frac{1 + x_3}{-x_1 + ix_2} = -\frac{1 + x_3}{x_1 - ix_2} .$$

Thus, the antihomography corresponding to the inversion has the form

$$t \to t' = -1/\bar{t} .$$

Since an arbitrary improper rotation can be expressed as a composition of a proper rotation and the inversion, it is easy to verify that improper rotations are represented by antihomographies. ⊠

Relation to Pauli matrices
The unitary matrix of the form used in Eq.(2.36) can be expressed as

$$\begin{bmatrix} q^0 + iq^3 & q^2 - iq^1 \\ -q^2 - iq^1 & q^0 - iq^3 \end{bmatrix} = q^\mu \varsigma_\mu ,$$

where

$$\varsigma_0 = \begin{bmatrix} 1 & 0 \\ 0 & 1 \end{bmatrix} , \quad \varsigma_1 = \begin{bmatrix} 0 & -i \\ -i & 0 \end{bmatrix} , \quad \varsigma_2 = \begin{bmatrix} 0 & 1 \\ -1 & 0 \end{bmatrix} , \quad \varsigma_3 = \begin{bmatrix} i & 0 \\ 0 & -i \end{bmatrix} .$$

For $i, j = 1, 2, 3$, these matrices satisfy $\varsigma_{(i)}\varsigma_{(i)} = -\varsigma_0$ (no summation), $\varsigma_i\varsigma_j = -\varsigma_j\varsigma_i$, $(i \neq j)$, and there occurs

$$\varsigma_1\varsigma_2 = \varsigma_3 , \quad \varsigma_2\varsigma_3 = \varsigma_1 , \quad \varsigma_3\varsigma_1 = \varsigma_2 .$$

Let $\tau := \begin{bmatrix} 0 & 1 \\ 1 & 0 \end{bmatrix} = \tau^{-1}$. The hermitian matrices $\sigma_k = i\tau^{-1}\varsigma_k\tau$ are called Pauli matrices. They are well known to physicists, especially to those working in particle physics. For the combination $q^\mu\varsigma_\mu$ to be a special unitary matrix, the real coefficients q^μ must satisfy (2.35). $\boxtimes$

2.6 Quaternions

The algebra of quaternions was formally defined in section 1.5.2 on Clifford algebras. This part is to provide more information on their properties. It is assumed that e_μ, $\{\mu = 0, 1, 2, 3\}$ constitute an orthonormal basis of a four dimensional vector space over real numbers. Besides the standard operations of the vector space, the internal multiplication of vectors is defined by the relations

$$e_\mu \diamond e_0 = e_0 \diamond e_\mu = e_\mu , \quad e_i \diamond e_j = \varepsilon_{ijk}e_k - \delta_{ij}e_0 ; \tag{2.39}$$

cf. Table (1.19). Hence, if $x = x^\mu e_\mu$, $y = y^\mu e_\mu$, where x^μ, y^μ are real numbers, the quaternion product $x \diamond y$ has the form

$$x \diamond y = \left(x^0 y^0 - x^i y^i\right) e_0 + \left(x^0 y^k + y^0 x^k + \varepsilon_{ijk}x^i y^j\right) e_k . \tag{2.40}$$

The quaternion e_0 is the identity of the above operation, i.e., $e_0 \diamond x = x = x \diamond e_0$. The quaternion algebra is associative but not commutative, i.e., in general $x \diamond y \neq y \diamond x$. The 'scalar parts' (coefficients of e_0) of $x \diamond y$ and $y \diamond x$ are equal. If vector parts of x and y ($x^i e_i$ and $y^i e_i$, respectively) are parallel vectors, the quaternions x and y do commute. The quaternion $x^* := x^0 e_0 - x^i e_i$ is called conjugate to $x = x^\mu e_\mu = x^0 e_0 + x^i e_i$. The norm $|x|$ of the quaternion x is defined by $|x| = \sqrt{x^\mu x^\mu}$. In the case when $|x| \neq 0$, due to $x \diamond x^* = |x|^2 e_0$,

$$x^{-1} = x^* / |x|^2 .$$

If $|x| = 0$, then $x = 0$; thus, the inverse of x does not exist only if $x = 0$. For non–zero x and y, because $x \diamond y \diamond y^{-1} \diamond x^{-1} = e_0$, there occurs $(x \diamond y)^{-1} = y^{-1} \diamond x^{-1}$. A quaternion x for which $|x| = 1$ is referred to as a unit (magnitude) quaternion; if x is a unit quaternion, then $x \diamond x^* = e_0$ and $x^\mu x^\mu = 1$. It is easy to show using (2.40) that the product of unit quaternions is a unit quaternion.

Matrices representing quaternions
Let the 4×4 real matrix $[x]^+$ associated with the quaternion x be defined by $[x]^+_{\mu\nu} = e_\mu \cdot (x \diamond e_\nu)$, where the dot $\cdot$ denotes the scalar product of two vectors. In a more explicit form

$$[x]^+ = \begin{bmatrix} x^0 & -x^1 & -x^2 & -x^3 \\ x^1 & x^0 & -x^3 & x^2 \\ x^2 & x^3 & x^0 & -x^1 \\ x^3 & -x^2 & x^1 & x^0 \end{bmatrix} .$$

The matrices $[x]^+$ and $[y]^+$ associated with quaternions x and y, respectively, satisfy

$$[x]^+[y]^+ = [x \diamond y]^+ \quad \text{and} \quad [x]^+ y = x \diamond y .$$

(In the last relation, the quaternions y and $x \diamond y$ are treated as 4×1 matrices.) Similarily, for the matrix $[x]^-$ defined by $[x]^-_{\mu\nu} = (e_\mu \diamond x) \cdot e_\nu$, one has $[x]^-[y]^- = [x \diamond y]^-$ and $[x]^- y^* = (x \diamond y)^*$. Additionally, there occurs

$$[x^*]^+ = ([x]^+)^T \quad \text{and} \quad [x^*]^- = ([x]^-)^T .$$

Moreover, the matrices satisfy $\det([x]^\pm) = |x|^4$ and

$$([x]^+)^T[x]^+ = |x|^2 I_4 \quad \text{and} \quad ([x]^-)^T[x]^- = |x|^2 I_4 .$$

These relations show that, if x is a unit quaternion, the matrices $[x]^\pm$ are special orthogonal. $\boxtimes$

The relationship between quaternions and rotations was explained in section 1.5.2 on Clifford algebra. The link also comes from the comparison of the relation (2.38) for the composition of rotations expressed through parameters q^μ and the relation (2.40) for the composition of quaternions.[2] They are identical. Taking into account the condition (2.35) ($q^\mu q^\mu = 1$), the unit quaternions correspond to rotations: if the quaternions q, q' correspond to orthogonal matrices O, O', then the quaternion product $q' \diamond q$ corresponds to $O'O$. This was already expressed in formal language by saying that the group of unit quaternions is homomorphic to $SO(3)$. Quaternion components as parameters of rotations are sometimes called Euler or Euler–Rodrigues parameters.

The relation between quaternion components and matrix components can be obtained from (2.34). With a and b expressed through q^μ,

$$O = \begin{bmatrix} (q^0)^2 + (q^1)^2 - (q^2)^2 - (q^3)^2 & 2(q^1 q^2 - q^0 q^3) \\ 2(q^1 q^2 + q^0 q^3) & (q^0)^2 - (q^1)^2 + (q^2)^2 - (q^3)^2 \\ 2(q^1 q^3 - q^0 q^2) & 2(q^2 q^3 + q^0 q^1) \end{bmatrix}$$

$$\begin{matrix} 2(q^1 q^3 + q^0 q^2) \\ 2(q^2 q^3 - q^0 q^1) \\ (q^0)^2 - (q^1)^2 - (q^2)^2 + (q^3)^2 \end{matrix} \Bigg] .$$

In brief notation, this takes the form

$$O_{ij} = \left((q^0)^2 - q^k q^k\right) \delta_{ij} + 2q^i q^j - 2\varepsilon_{ijk} q^0 q^k . \tag{2.41}$$

It is noteworthy that the matrix O is part of the product of the matrices $[q]^-$ and $[q]^+$

$$[q]^-[q]^+ = \begin{bmatrix} 1 & 0 \\ 0 & O \end{bmatrix} . \tag{2.42}$$

Since

$$[q]^-_{i\mu} = (\delta_{i\mu}\delta_{0\nu} - \delta_{0\mu}\delta_{i\nu} - \varepsilon_{imn}\delta_{m\mu}\delta_{n\nu}) q^\nu ,$$

---

[2] Alternatively, compare the multiplication rules for ς_μ matrices with 2.39.

$$[q]^+_{\mu i} = (\delta_{i\mu}\delta_{0\nu} - \delta_{0\mu}\delta_{i\nu} + \varepsilon_{imn}\delta_{m\mu}\delta_{n\nu})\, q^\nu \ ,$$

the relationship between a special orthogonal matrix O and corresponding unit quaternions can also be written as

$$O_{ij} = [q]^-_{i\kappa}[q]^+_{\kappa j} = \mathsf{G}_{ij\mu\nu} q^\mu q^\nu \ , \tag{2.43}$$

where $\mathsf{G}_{ij\mu\nu}$ is defined by

$$\mathsf{G}_{ij\mu\nu} = \delta_{ij}(\delta_{0\mu}\delta_{0\nu} - \delta_{k\mu}\delta_{k\nu}) + \delta_{i\mu}\delta_{j\nu} + \delta_{i\nu}\delta_{j\mu} - \varepsilon_{ijk}(\delta_{k\mu}\delta_{0\nu} + \delta_{k\nu}\delta_{0\mu}) \ . \tag{2.44}$$

This can be verified by comparing individual components with 2.41. Conversely, for a given O, the components $\pm q^\mu$ can be extricated from the formula $q^\mu q^\nu = (\delta_{\mu\nu} + \mathsf{G}_{ij\mu\nu}O_{ij})/4$, which follows from Eq.(2.43) and the identity $\mathsf{G}_{ij\iota\kappa}\mathsf{G}_{ij\mu\nu} = 2(\delta_{\iota\mu}\delta_{\kappa\nu} + \delta_{\iota\nu}\delta_{\kappa\mu}) - \delta_{\iota\kappa}\delta_{\mu\nu}$.

The trace of O is determined by the scalar part of the corresponding quaternion

$$O_{ii} = 4(q^0)^2 - 1 \ , \quad \text{and} \quad q^0 = \pm\frac{(O_{ii}+1)^{1/2}}{2} \ .$$

The vector part of the quaternion is related to the antisymmetric part of O; equation (2.41) leads to

$$\varepsilon_{ijk}O_{jk} = -4q^0 q^i \ .$$

For $q^0 \neq 0$, which is equivalent to $O_{ii} \neq -1$, we have

$$q^i = \mp\frac{\varepsilon_{ijk}O_{jk}}{2(O_{ll}+1)^{1/2}} \ . \tag{2.45}$$

The choice of the sign must be coordinated in all components of the quaternion. When $q^0 = 0$ then $O_{ij} = 2q^i q^j - \delta_{ij}$.

Using the above formulas, the quaternion components can be easily related to axis and angle parameters. The direction of the rotation axis is expressed via quaternion components by

$$n_i = \frac{q^0 q^i}{\sqrt{q^0 q^0 \ q^k q^k}} \quad \text{for} \quad q^0 q^k \neq 0 \quad \text{and} \quad n_i = q^i \quad \text{for} \quad q^0 = 0 \ .$$

The inverse formulas are also simple. The relation (2.18) between the trace of the rotation matrix and the rotation angle ω gives $q^0 = \pm\, |\cos(\omega/2)|$. The comparison of the expressions for the rotation axis (2.19) and (2.45) leads to the relation $q^i = \pm\sqrt{3 - O_{jk}O_{kj}}/(2(O_{ll}+1)^{1/2})n_i$, which can be transformed to $q^i = \pm\, |\sin(\omega/2)|\, n_i$. With ω in the range $[0, \pi]$, the values of $\sin(\omega/2)$ and $\cos(\omega/2)$ are non-negative and the quaternion components can be expressed as

$$q^0 = \pm\cos(\omega/2) \quad \text{and} \quad q^i = \pm\sin(\omega/2)n_i \ . \tag{2.46}$$

In this case, the relationship between unit quaternions to the parameters r^i of section 2.2 has the form

$$r^i = q^i/q^0 \ , \quad q^0 \neq 0 \ . \tag{2.47}$$

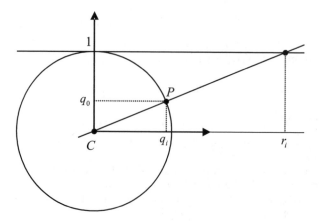

Fig. 2.3. Gnomic projection.

Unit quaternions are points on a unit sphere in four dimensional space. Opposite points of the sphere correspond to the same rotation. The relation (2.47) above is the gnomic projection of the hemisphere on the three dimensional hyperplane tangent to the sphere at the point $q^0 = 1$.

Gnomic projection
In the gnomic projection of the hemisphere (with the center at C) on a plane tangent to the sphere, the image of a given point P of the hemisphere is the point of intersection of the line CP with the plane (Fig. 2.3). ⊠

Orthographic projection
With $q^0 \geq 0$, the quaternion vector part q^i alone can be used for determining orientations. I.e., the 'orthographic projection' of the unit quaternion hemisphere is used in this case. The scalar part, if needed, is calculated from $q^0 = \sqrt{1 - q^i q^i}$. ⊠

Representation by non–zero quaternions
As H.Schaeben pointed out to us, proper rotations can be represented by general non–zero quaternions without the restriction on their magnitude. The rotation corresponding to a non–zero quaternion q is the same as the one ascribed above to the unit quaternion $q/|q|$. If q' and q'' are two non–zero quaternions then $q'' \diamond q'$ is nonzero and corresponds to the rotation being composition of rotations corresponding to q' and q''. This means that the composition of rotations performed via multiplication of quaternions does not require their normalization. In practice, it is convenient and routine to represent rotations by quaternions of unit magnitude. ⊠

Dirac's construction
Let us forget about (2.46) for a while and express the unit quaternion as

$$q^0 = \cos(\omega/2) \ , \quad q^i = \sin(\omega/2)n_i \ . \tag{2.48}$$

The change of parameters from (n, ω) to $(n, \omega + 2\pi)$ causes the change from q to $-q$. In order to restore q, the angle ω must be increased by 4π. As q and $-q$ represent the same rotation, the rotation about n by 2π does not change the initial configuration.

However, in quantum mechanical interpretation of angular momentum, a distinction is made between rotations by 2π and 4π. There is also a macroscopic construction devised by P.Dirac in which rotations by $2k\pi$ differ from those by $4k\pi$ (k – an integer). Imagine an object attached to its surrounding by strings. After 4π rotation of the object, the strings can be disentangled without cutting; see Fig. 2.4. That is impossible if the rotation is by 2π. $\boxtimes$

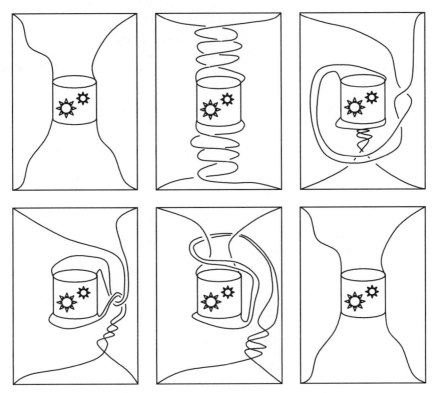

Fig. 2.4. Dirac's construction. The object with attached strings (first picture) is rotated by 4π (second picture). The remaining pictures show how the strings are disentangled.

Analogy to complex numbers

In association to unit quaternions and their relation to proper rotations in 3 dimensional space, it is worth recalling complex numbers of modulus 1 and their relation to rotations in 2 dimensional space; in the latter case, the correspondence is one-to-one.

Based on (2.48), a unit quaternion can be expressed as $\cos(\theta) + \iota \sin(\theta)$, where $\theta = \omega/2$ and ι is a unit quaternion with components $(0, n_1, n_2, n_3)$. As $\iota\iota = -1$, unit quaternions with vanishing scalar parts play the role similar to the complex number $i := \sqrt{-1}$. The analogy does not go much further since ι is not unique. (It is easy to see that such quaternions correspond to half–turns.) Similarities to complex numbers appear if ι is fixed. For example, the analogue of the de Moivre formula is true: $(\cos(\theta) + \iota \sin(\theta))^k = \cos(k\theta) + \iota \sin(k\theta)$. $\boxtimes$

Quaternion components $\leftrightarrow$ Euler angles
Using (2.33), the quaternion components can be expressed through Euler angles

$$q^0 = \cos(\phi/2)\cos((\varphi_1 + \varphi_2)/2) \,, \quad q^1 = -\sin(\phi/2)\cos((\varphi_1 - \varphi_2)/2) \,,$$
$$q^2 = -\sin(\phi/2)\sin((\varphi_1 - \varphi_2)/2) \,, \quad q^3 = \cos(\phi/2)\sin((\varphi_1 + \varphi_2)/2) \,.$$

Again, the quaternion with the opposite sign represents the same rotation. From these relations we obtain the inverse formulas for the calculation of the Euler angles from the quaternion. Let $\chi := [((q^0)^2 + (q^3)^2)((q^1)^2 + (q^2)^2)]^{1/2} = (\sin \phi)/2$. If $\chi \neq 0$,

$$\cos \phi = ((q^0)^2 + (q^3)^2) - ((q^1)^2 + (q^2)^2) \,,$$
$$\cos \varphi_1 = (-q^0 q^1 + q^2 q^3)/\chi \,, \qquad \sin \varphi_1 = (-q^0 q^2 - q^1 q^3)/\chi \,,$$
$$\cos \varphi_2 = (-q^0 q^1 - q^2 q^3)/\chi \,, \qquad \sin \varphi_2 = (-q^0 q^2 + q^1 q^3)/\chi \,.$$

If $\chi = 0$, there occurs $q^1 = 0 = q^2$ or $q^0 = 0 = q^3$. In the first case, we have

$$\phi = 0 \,, \quad \cos(\varphi_1 + \varphi_2) = (q^0)^2 - (q^3)^2 \,, \quad \sin(\varphi_1 + \varphi_2) = 2q^0 q^3 \,.$$

In the second one, the angles are given by

$$\phi = \pi \,, \quad \cos(\varphi_1 - \varphi_2) = (q^1)^2 - (q^2)^2 \,, \quad \sin(\varphi_1 - \varphi_2) = 2q^1 q^2 \,.$$

Both instances correspond to singularities of Euler angles and only combinations of φ_1 and φ_2 are determinable. $\boxtimes$

Quaternionic form of transformation of a vector
Based on (2.41), the transformation of Cartesian coordinates $x_i' = O_{ij}x_j$ can be expressed as

$$x_i' = \left((q^0)^2 - q^k q^k\right) x_i + 2q^i q^j x_j - 2\varepsilon_{ijk} q^0 x_j q^k \,.$$

The same form has the i-th quaternion component of the product qxq^* where x is the "vector" quaternion $x = 0e_0 + x_1 e_1 + x_2 e_2 + x_3 e_3$. Thus, with $x' = x_i' e_i$ $(i = 1, 2, 3)$, the transformation can be written as

$$x' = q \diamond x \diamond q^{-1} = q \diamond x \diamond q^* \,. \tag{2.49}$$

This rule follows also from the interpretation of quaternions as elements of the Clifford algebra $\mathcal{C}_3^+$ and (1.20).

The isomorphism between the group of unit quaternions and $SU(2)$ and the fact that the quaternions e_μ correspond to ς_μ matrices allows us to write

$$X' = UXU^\dagger \,,$$

where the unitary matrix U is related to q by (2.36), $X = x_i \varsigma_i$ and $X' = x_i' \varsigma_i$. These vector transformation rules are similar to Eq.(2.16). $\boxtimes$

Improper rotations and quaternions
One may ask about the possibility to represent improper rotations by quaternions. With quaternion algebra seen as a standing alone construct, it is natural to

follow an approach due to Coxeter (1946): If quaternions are selected to represent all rotations, to each quaternion a flag must be assigned to indicate whether it represents a proper or improper rotation. Below, the role of the flag is played by the 'prime' symbol. It was already said that an improper rotation can be expressed by a composition of a proper rotation and inversion. Thus, based on Eq.(2.49), the transformation of the type

$$x' = -q' \diamond x \diamond q'^* \tag{2.50}$$

corresponds to an improper rotation of a vector. If this transformation is a reflection, the points x on the reflection plane satisfy $x = -q' \diamond x \diamond q'^*$ or $x \diamond q' + q' \diamond x = 0$. The scalar part of $x \diamond q' + q' \diamond x$ is equal to $-2q'^i x_i$; hence, $q'^i x_i = 0$, i.e., the vector part of the quaternion corresponding to a reflection is perpendicular to the reflection plane. The i-th component of the vector part of $x \diamond q' + q' \diamond x$ is equal to $2q'^0 x_i$; hence, $q'^0 = 0$, i.e., the scalar part of the quaternion corresponding to a reflection vanishes. This is in agreement with the already made statement (section 2.3) that a reflection can be seen as the composition of the inversion and the rotation about an axis perpendicular to the reflection plane by the angle of π.

However, the account given above can be seen in a different and more transparent light after recalling the Clifford algebra $\mathcal{C}_3$ and the groups $Pin(3)$ and $Spin(3)$ of section 1.5.2. The truly natural way is to portray quaternions (of $Spin(3)$) representing proper rotations as quantities generated by elements of $Pin(3)$ representing reflections. ⊠

⊡ *More on factorizations*

With $q^0 \neq 0$, the factorization (2.42) of a special orthogonal matrix is directly related to the Cayley factorization (2.3). This is visible based on the expressions

$$[q]^- = q^0 \begin{bmatrix} 1 & -r^T \\ r & I - R \end{bmatrix} \quad \text{and} \quad [q]^+ = \left(([q]^+)^T \right)^{-1} = \frac{1}{q^0} \begin{bmatrix} 1 & r^T \\ -r & I + R \end{bmatrix}^{-1} ,$$

which follow from (2.47), (2.12), and from the orthogonality of $[q]^+$.

Additionally, we will consider the factorization (2.42) in relation to the decomposition of a proper rotation into two reflections. Let O be a composition of the reflections O_1 and O_2 ($O = O_2 O_1$), and the unit quaternions q'_1 and q'_2 correspond to O_1 and O_2, respectively. We have

$$[q]^- [q]^+ = \begin{bmatrix} 1 & 0 \\ 0 & O \end{bmatrix} = \begin{bmatrix} 1 & 0 \\ 0 & -O_2 \end{bmatrix} \begin{bmatrix} 1 & 0 \\ 0 & -O_1 \end{bmatrix} = [q'_2]^- [q'_2]^+ [q'_1]^- [q'_1]^+ .$$

Since $q'^0_1 = 0$ and $q'^0_2 = 0$, the matrices $[q'_1]^-$ and $[q'_2]^+$ commute, and $[q]^- [q]^+ = [q'_2]^- [q'_1]^- [q'_2]^+ [q'_1]^+ = [q'_2 \diamond q'_1]^- [q'_2 \diamond q'_1]^+$. This leads to the relation

$$q = \pm q'_2 \diamond q'_1 ,$$

which means that the unit quaternion q corresponding to O is a product of two unit quaternions with zero scalar components. Again, this statement becomes obvious when considered from the viewpoint of the Clifford algebra $\mathcal{C}_3$. ⊠

2.7 Rotation Vector

† Rodrigues parameters and the vector part of the quaternion are proportional to the Cartesian components of the unit vector of the rotation axis n. Let

us define general parameters of the form $\rho^i = f(\omega)n_i$, where the function $f : [0, \pi] \to R$ is assumed to be strictly monotone with $f(0) = 0$ (Fig. 2.5).

If $\omega = 0$, then $\rho^i = 0$. The magnitude of a general rotation "vector" ρ^i is determined by f

$$\sqrt{\rho^i \rho^i} = |f(\omega)| \ .$$

The manifold of proper rotations is covered if the rotation vector ρ is in the "parametric ball" $\sqrt{\rho^i \rho^i} \leq |f(\pi)|$. When $\omega = \pi$, the triplets $\{\rho^i\}$ and $\{-\rho^i\}$ represent the same rotation.

As ρ^i is proportional to n_i, we have

$$\rho^i \propto \varepsilon_{ijk} O_{jk} \ ,$$

with the proportionality coefficient depending on f. The orthogonal matrix can be expressed by ρ^i using the Rodrigues formula (2.21). Its O_{ij} entry is a combination of δ_{ij}, $\rho^i \rho^j$ and $\varepsilon_{ijk}\rho^k$; again the coefficients depend on $f(\omega)$. To write down the explicit form of that relation, ω must be given in terms of $\sqrt{\rho^i \rho^i}$, and a difficulty may arise if the function f^{-1} inverse to f is complicated.

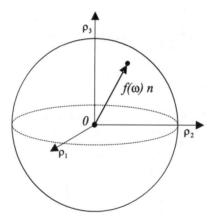

Fig. 2.5. Graphic representation of the parameters ρ^i.

The function f can be specified according to particular requirements. For Rodrigues parameters and the vector part of a quaternion (of upper hemisphere), f is given by $\tan(\omega/2)$ and $\sin(\omega/2)$, respectively; cf. Eqs. (2.20) and (2.46). See Fig. 2.6 for graphs.

In a narrow sense, the rotation vector is defined as the set of parameters ω^i given by the relation

$$\omega^i = \omega n_i \ , \tag{2.51}$$

i.e., in this case $f(\omega) = \omega$.

Let us also mention the parameters ζ^i defined as

$$\zeta^i = n_i \tan(\omega/4) \ . \tag{2.52}$$

They are sometimes referred to as modified Rodrigues parameters. The metric tensor, which will be considered in chapter 3, has a very simple form in this parameterization. The modified Rodrigues parameters are related to quaternions via the stereographic projection of the quaternion upper hemisphere on R^3.

Stereographic projection in 4D
The stereographic projection of a point $q = (q^0, q^1, q^2, q^3)$ of the quaternion upper hemisphere ($q^0 \geq 0$) on R^3 is the point of intersection of the line through q and the "south pole" of the sphere $s = (-1, 0, 0, 0)$ with the hyperplane $(0, \zeta^1, \zeta^2, \zeta^3)$. A point on the line can be expressed as $s + \lambda(q - s)$, where λ is a real number. Hence,

$$(0, \zeta^1, \zeta^2, \zeta^3) = (-1, 0, 0, 0) + \lambda((q^0, q^1, q^2, q^3) - (-1, 0, 0, 0)) .$$

The first of these four equations gives $\lambda = 1/(1 + q^0)$, and the remaining relations lead to

$$\zeta^i = q^i/(1 + q^0) . \tag{2.53}$$

From (2.46), one gets $\zeta^i = \tan(\omega/4)n_i$.
Other parameters of similar kind can be constructed by taking projections other than gnomic or stereographic. E.g., for the interpretation of parameters

$$\xi^i = q^i/(a + q^0) , \quad 0 \leq a \leq 1$$

see (Schaub & Junkins, 1996). With $a = 0$ ($a = 1$), the parameters ξ^i are reduced to the Rodrigues (modified Rodrigues) parameters. The requirement $q^0 \geq 0$ becomes unnecessary if the above expression for ξ^i is replaced by $\xi^i = q^i/(\text{sign}(q^0)a + q^0)$. ⊠

The modified Rodrigues parameters are related to the anitisymmetric matrix of the second order Cayley transformation (2.17); the parameters ζ^i of $O = (I - R)^2(I + R)^{-2}$ are linked to R via $\zeta^i = \varepsilon_{ijk}R_{jk}/2$ (Tsiotras *et al.*, 1997). More generally, the orthogonal matrix O corresponding to the rotation vector $n_i \tan(\omega/(2m))$ is given by m–th order Cayley transformation $O = (I - R)^m(I + R)^{-m}$, where $R_{ij} = \varepsilon_{ijk}n_i \tan(\omega/(2m))$. For $m = 1$ and $m = 2$, we get the particular cases associated with the Rodrigues parameters and the modified Rodrigues parameters, respectively.
Besides the already listed cases, the so–called "isochoric" parameters with

$$f(\omega) = \left(\frac{3}{4\pi^2}(\omega - \sin \omega) \right)^{1/3} \tag{2.54}$$

are of interest (Frank, 1988). In this parameterization, the determinant of the metric tensor matrix is constant. In analogy to 'equal-area projection' of the sphere S^2 on a plane, this parameterization can be called 'equal-volume projection' of the sphere S^3 on three dimensional flat space.

Equal-volume projection
In the 'equal-volume projection', the ratio: (volume of a region on S^3) to (volume of the region's image) is constant for all regions. The isochoric parameters are obtained by projecting the upper ($q^0 \geq 0$) quaternion hemisphere. In 'equal-volume projection' the image h_i of a point q^μ on the hemisphere is given by

$$h_i \propto \frac{q^i}{\sqrt{q^j q^j}} \left(\arccos q^0 - q^0 \sqrt{q^k q^k} \right)^{1/3} ,$$

and this means that $h_i \propto f(\omega)n_i$, where f is given by Eq.(2.54). $\boxtimes$

Parameterization of neighborhood of identity
In order to cover all proper rotations, the function f was assumed to be strictly monotone on $[0, \pi]$. This assumption can be modified if someone is interested in a map of the neighborhood of the identity. E.g., the parameterization by

$$p^i = -\sin(\omega)n_i$$

can be used. The numbers p^i for ω are the same as for $\pi - \omega$, i.e., they provide a one–to–one correspondence only for rotations by angles not larger than $\pi/2$. Their relation to the orthogonal matrix is very simple $p^i = \varepsilon_{ijk}O_{jk}/2$, and the antisymmetric matrix corresponding to p^i is the antisymmetric part of O

$$\varepsilon_{ijk}p^k = (O_{ij} - O_{ji})/2 \ .$$

In literature, one may encounter the set of parameters p^μ, $\mu = 0, 1, 2, 3$ with $p^0 = \cos(\omega)$; the additional quantity p^0 serves as a flag which allows rotations by angles beyond $\pi/2$ to be considered. $\boxtimes$

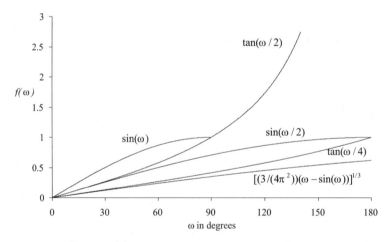

Fig. 2.6. Graphs of functions used as magnitudes of rotation vectors.

2.8 Rotation Matrix in Non–Cartesian Coordinate Systems

Let us take a closer look at the rotation matrices in the systems in which coordinate axes are non–coplanar straight lines but not necessarily orthogonal. Let the vectors $\{e_i,\ i = 1, 2, 3\}$ and $\{a_i,\ i = 1, 2, 3\}$ constitute bases of such coordinate systems. Vectors of one base can be expressed as combinations of vectors of the other base, i.e.,

$$a_i = A^j{}_i e_j \ , \quad e_i = B^j{}_i a_j \ .$$

Thus, $e_i = B^k{}_i A^j{}_k e_j$ and $a_i = A^k{}_i B^j{}_k a_j$. Hence, A and B are mutually inverse matrices

$$A^i{}_k B^k{}_j = \delta^i_j \ , \quad B^i{}_k A^k{}_j = \delta^i_j \ .$$

Relations between vector components x^i_e and x^i_a of the vector x in the two bases can be easily obtained from $x = x^i_e e_i = x^i_a a_i$ and $x^i_a a_i = B^i{}_j x^j_e a_i$, $x^i_e e_i = A^i{}_j x^j_a e_i$; this leads to

$$x^i_a = B^i{}_j x^j_e \ , \quad x^i_e = A^i{}_j x^j_a \ .$$

More generally, components of a pseudotensor t of weight w are transformed between the coordinate systems as

$$t_e{}^{ij...k}{}_{lm...n} = (\det(A))^{-w} t_a{}^{pq...r}{}_{st...u} A^i{}_p A^j{}_q ... A^k{}_r B^s{}_l B^t{}_m B^u{}_n \ . \qquad (2.55)$$

For instance, due to (2.9), the permutation symbol ε represents pseudotensor of weight 1. If $w = 0$, then t is simply a tensor. A tensor is called contravariant if it is transformed using $A^i{}_j$, and is covariant if the transformation involves $B^i{}_j$. The difference in the transformation rule is reflected in the level of the indices: upper indices for contravariant tensors, lower indices for covariant tensors.

The matrix g with entries given by $g_{ij} = a_i \cdot a_j$ is the metric tensor in the a_i basis. The contravariant form of the metric tensor g^{ij} is the inverse of $g = \{g_{ij}\}$. The symbol g will denote the matrix $\{g_{jk}\}$, i.e. the one with lower indices. With v^j being a contravariant vector, the set $g_{ij}v^j$ obeys the covariant rule of transformation. Since the metric tensor is unique, this is a method of lowering indices. Analogously, vector or tensor indices are rised using g^{ij}.

If the system based on $\{e_i, \ i = 1, 2, 3\}$ is Cartesian, then $g_{ij} = A^k{}_i A^l{}_j (e_k \cdot e_l) = A^k{}_i A^l{}_j \delta_{kl}$ and $g^{ij} = B^i{}_k B^j{}_l \delta^{kl}$. In the Cartesian coordinate system, the rotation represented by an orthogonal matrix O gives the change of vector coordinates which can be expressed by the relationship $x'^i_e = O^i{}_j x^j_e$. In the non–Cartesian system, the transformation has the form

$$x'^i_a = R^i{}_j x^j_a \ ,$$

with

$$R^i{}_j = B^i{}_k O^k{}_l A^l{}_j \quad \text{and} \quad O^i{}_j = A^i{}_k R^k{}_l B^l{}_j \ . \qquad (2.56)$$

The orthogonality condition $O^k{}_i O^k{}_j = \delta_{ij}$ leads to $A^k{}_m R^m{}_l B^l{}_i A^k{}_s R^s{}_l B^l{}_j = \delta_{ij}$ or

$$g_{ij} R^i{}_k R^j{}_l = g_{kl} \ ; \qquad (2.57)$$

cf. Eq.(1.8). In analogous way, the condition $R^k{}_i R^l{}_j g^{ij} = g^{kl}$ can be obtained from $O^i{}_k O^j{}_k = \delta^{ij}$.

Generally, relations given earlier for Cartesian systems, can be expressed in non–orthonormal coordinate systems. For example, the Rodrigues formula Eq.(2.21) can be transformed to

$$R^i{}_j = \cos\omega\,\delta^i{}_j + (1 - \cos\omega)g_{jk}n^i n^k / \det(g) - \sin\omega\varepsilon_{jkl}g^{ik}n^l \;, \qquad (2.58)$$

where R is related to the orthogonal matrix via (2.56) and n^i are components of n in the non–orthonormal coordinate system. The relation between n^i and the components n^i_e in the Cartesian system is

$$n^i_e = (\det(A))^{-1}A^i{}_j n^j \;,$$

i.e., the components of n are transformed as a pseudotensor of weight 1. Also, $\varepsilon_{ijk}A^i{}_s A^k{}_t A^j{}_u = \det(A)\varepsilon_{stu}$ is needed for the derivation of (2.58). Direct calculation shows that the trace of R is

$$R^i{}_i = 1 + 2\cos\omega \;; \qquad (2.59)$$

i.e., the expression is similar to that in Cartesian systems.

□ *Gibbs vectors in non–Cartesian systems*
    Later on, we will need analogues of the Rodrigues (Gibbs) vector and the relation (2.13) in a non–Cartesian coordinate system. Based on (2.13), there occurs $r^i = -\varepsilon^{ijk}\delta_{js}A^s{}_m R^m{}_l B^l{}_k / (1 + R^u{}_u)$. Hence, by using $\varepsilon^{stu}A^i{}_s A^j{}_t A^k{}_u = \det(A)\varepsilon^{ijk}$, we have $\det(A)B^s{}_i r^i = -\varepsilon^{stl}A^j{}_t A^j{}_m R^m{}_l / (1 + R^u{}_u)$ or

$$\varrho^s = -\frac{1}{1 + R^u{}_u}\varepsilon^{stl}g_{tm}R^m{}_l \;, \qquad (2.60)$$

where

$$\varrho^s = \det(A)B^s{}_i r^i \;.$$

By this definition $g_{st}\varrho^s\varrho^t / \det(g) = \delta_{kl}r^k r^l$. The parameters ϱ^s determine a rotation in the system with the metric g. The relation (2.60) allows ϱ to be calculated from R. Also R can be expressed via ϱ. Steps similar to those used above lead to

$$R^k{}_l = \frac{1}{\det(g) + g_{st}\varrho^s\varrho^t}((\det(g) - g_{uw}\varrho^u\varrho^w)\delta^k{}_l + 2g_{tl}\varrho^k\varrho^t - 2\det(g)\varepsilon_{lnm}g^{mk}\varrho^n) \;.$$
$$(2.61)$$

This formula is valid for non–Cartesian coordinate systems and Eq.(2.14) is its particular form corresponding to $g_{ij} = \delta_{ij}$. ⊠

2.9 Miller Indices

Crystal structures are built on space lattices. Lattice points (ends of lattice vectors) are "decorated" with atoms in specific arrangements. For a given material, the arrangements of atoms are identical at each lattice point.

Let a_i $(i = 1, 2, 3)$ be three linearly independent vectors. Three dimensional lattice is a set of vectors determined by linear combinations of the form $u^i a_i$ with integer u^i. In agreement with the general account given in section 2.8, the *crystallographic* metric g is defined as $g_{ij} = a_i \cdot a_j$, and the contravariant form $\{g^{ij}\}$ of g is determined by the conditions $g^{ij}g_{jk} = \delta^i{}_k$.

Contravariant coordinates of lattice vectors (i.e., coordinates with integer values) indicate directions through lattice points, and are called (Miller) direction indices. Covariant coordinates with co–prime integer values are used for specifying crystallographic planes; the coordinates are called Miller (plane) indices. For (non–zero) Miller indices h_i^*, the vector $h_i^* g^{ij} a_j$ is perpendicular to the plane through the points $a_{(i)}/h_{(i)}^*$, (no summation, $i = 1, 2, 3$). Vectors $a^{*i} := g^{ij} a_j$ constitute a basis of the lattice referred to as *reciprocal* to the lattice based on a_i. The vectors a^{*i} satisfy the relationship $a^{*i} \cdot a_j = \delta^i{}_j$. A vector of the reciprocal lattice is given by $h_j^* a^{*j}$. Direction and plane indices are denoted by $[u^1 u^2 u^3]$ and $(h_1^* h_2^* h_3^*)$, respectively. Since $(h_j^* a^{*j}) \cdot (u^i a_i) = u^i h_i^*$, indices $[u^1 u^2 u^3]$ of directions parallel to the plane $(h_1^* h_2^* h_3^*)$ satisfy (Weiss zone law) $u^i h_i^* = 0$.

It is a common practice in textures of metals to specify crystallite orientations using Miller indices. From the formal point of view, the Miller indices – as integers – cannot serve as a parameterization of a continuous manifold. Moreover, they are applicable only in some special cases. However, practical reasons are strong enough for the Miller indices to be a kind of standard for naming orientations. Therefore, understanding them is inevitable for everyone involved in crystallographic textures.

In principle, the method of using Miller indices as determinants of crystal orientations is simple: one specifies crystallographic indices of some fixed directions or fixed planes. In the most important standard of rolled metal sheets these are the plane of rolling and the direction of rolling; an orientation is presented as

$$(h \, k \, l)[u \, v \, w] \,,$$

which means that the crystallite is positioned in such a way that the rolling plane is parallel to the crystallographic plane $(h \, k \, l)$, and the rolling direction is parallel to the crystallographic direction $[u \, v \, w]$. By convention, the reference coordinate system attached to the sample is right–handed and has e_1 axis along the rolling direction and e_3 axis normal to the rolling plane.

In practice, general application of the method is more complicated and restricted because of symmetrically equivalent planes and directions present due to improper symmetry operations. This subject, however, will be discussed in section 6.3.

Typically, Miller indices are used for specifying crystallite orientations in metals belonging to the *cubic system*. The mutually perpendicular unit vectors along edges of the standard cubic unit cell are determinants of crystal orientation. If these vectors are named a_i ($i = 1, 2, 3$), and constitute right–handed system, the entry O_{ij} of the orthogonal matrix is the j-th component of the vector a_i in the reference coordinate system. In that case, the Miller indices of a given plane are the same as the indices of the direction normal to it. As the $[h \, k \, l]$ and $[u \, v \, w]$ are orthogonal, there occurs $hu + kv + lw = 0$. The relation between the indices and the orthogonal matrix follows from

$$\begin{bmatrix} h \\ k \\ l \end{bmatrix} \propto O \begin{bmatrix} 0 \\ 0 \\ 1 \end{bmatrix} \,, \quad \text{and} \quad \begin{bmatrix} u \\ v \\ w \end{bmatrix} \propto O \begin{bmatrix} 1 \\ 0 \\ 0 \end{bmatrix} \,.$$

Thus,

$$[O_{13} \, O_{23} \, O_{33}] \propto [h \, k \, l] \,, \quad [O_{11} \, O_{21} \, O_{31}] \propto [u \, v \, w] \,. \tag{2.62}$$

The indices are assumed to be integers. Therefore, they are calculated by multiplying the right–hand side of these relations by the largest acceptable value of an index and rounding the results to the nearest integers. The conditions following from the fact that O is special orthogonal, determine the second column of the matrix.

For other (i.e., *non–cubic*) crystallographic systems the appropriate expressions become more complicated. In general, the vectors determining the unit cell are not orthonormal. If crystal orientations are to be expressed by orthogonal matrices, there must be a set of three orthonormal vectors associated with the crystal; it is called Cartesian crystal coordinate system.

The first issue is to express the Cartesian coordinates (h_1^e, h_2^e, h_3^e) of the normal to a crystallographic plane by Miller indices of the plane. Let $A^j{}_i$ be the j-th coordinate of the i-th basis vector a_i of the direct lattice in the Cartesian crystal coordinate system. Consider a crystallographic plane with Miller indices $(h_1^* \ h_2^* \ h_3^*) = (h \ k \ l)$. Since the Miller indices of planes are vector covariant coordinates, the Cartesian coordinates h_i^e, $(i = 1, 2, 3)$ of the vector normal to that plane in the crystal coordinate system are related to the Miller indices h_i^* through

$$h_i^* = A^j{}_i h_j^e \ . \tag{2.63}$$

As for the Miller indices of directions $[u \ v \ w] = [u^1 \ u^2 \ u^3]$, they are related to the Cartesian coordinates (u_e^1, u_e^2, u_e^3) of the vector along that direction by

$$u^i = B^i{}_j u_e^j \ , \tag{2.64}$$

where B is inverse to A: $A^i{}_k B^k{}_j = \delta^i{}_j$. It is easy now to relate the indices to the entries of the rotation matrix

$$h_i^* = A^j{}_i O_{j3} \ , \quad \text{and} \quad u_i = B^i{}_j O_{j1} \ .$$

The sequence

$$hu + kv + lw = h_i^* u^i = A^j{}_i O_{j3} B^i{}_l O_{l1} = \delta_{13} = 0$$

shows that also in this general case the zone law $hu + kv + lw = 0$ is obayed.

For some crystal coordinate systems, more specific forms of the A matrix are in use. (See, e.g., Busing & Levy (1967), Kabsch (1977), Burenkov & Popov (1994).) Here, no constraints are imposed on the choice of the system or the unit cell. The cell may be non–conventional as long as the Miller indices are consistent with that cell.

☐ *Derivation of (2.63) and (2.64)*
The relations (2.63) and (2.64) given above can be easily justified using material of section 2.8. The matrix A relates the basis vectors of the direct lattice to the basis e_i of the Cartesian coordinate system

$$A^j{}_i = a_i \cdot e_j \ ,$$

A vector of the direct lattice is given by $u^i a_i = u_e^i e_i$. By taking the scalar product of both sides with e_j, one gets $u_e^j = A^j{}_i u^i$ and (2.64). Since for the Cartesian system

$e^{*j} = e_j$, after taking the scalar product of both sides of $h_j^* a^{*j} = h_j^e e^{*j}$ with a_i, we get $h_i^* = h_j^e A^j{}_i$. $\boxtimes$

> **Conventions for direction indices**
> Above and further, the 'standard' indices are used. If someone wants to use 'Weber' direction indices based on hexagonal unit cell, additional transformations must be applied. For Miller indices $(h\,k\,l)$ of a plane, the corresponding Miller–Bravais indices are $(h \quad k \quad -(h+k) \quad l)$. Miller indices of directions can be transformed to 4-tuples so the orthogonality relations are retained. For Miller indices $[u\,v\,w]$ of a direction, the corresponding 4-tuple indices are $[2u-v \quad 2v-u \quad -(u+v) \quad 3w]/3$. It is trivial to check that $(h\,k\,l) \cdot [u\,v\,w] = (h \quad k \quad -(h+k) \quad l) \cdot [2u-v \quad 2v-u \quad -(u+v) \quad 3w]/3$, where the $\cdot$ operation is defined as the sum of products of corresponding array elements. See, e.g., Niggli, 1928 (p.98) or Frank, 1965. $\boxtimes$

2.10 Computational Properties of Parameterizations

Analysis of textures often requires large scale computations. The question arises about most advantageous choice of parameters. Besides the conceptual suitability to a considered problem, one would like the parameters to guarantee robust, accurate and effective (in terms of the execution time) computations. Operation count and analytical estimation of numerical accuracy are immaterial for practical applications because of large variety of optimization tools used in contemporary hardware and software. Therefore, the simplest way to select the best parameterization for a given case is by empirical tests and comparisons.

Obviously, the performance depends not only on the environment but above all on the type of problem to be solved and applied algorithm. For example, the important simple task of composing rotations is best performed using parameterizations without singularities and involving relatively small number of algebraic operations and without trigonometric or other costly transcendental functions. In practice, this means that only the representations by orthogonal matrices or unit quaternions can be applied. The composition is slightly faster if performed using quaternions than by multiplication of matrices.[3] However, if large sequences of rotations are composed, the result is affected by errors (the matrix is no longer orthogonal or the quaternion's magnitude differs from 1) and must be projected on the subspace representing rotations. While for a quaternion this is done simply by normalization, orthonormalization of a matrix – i.e., finding the 'nearest' orthogonal matrix – is a costly process; see section 5.1.1.

Frequently, more important than effectiveness and accuracy is the clarity of problem formulation and transparency of applied equations. A parameterization very useful in one case may be completely inappropriate in another. Moreover, parameterizations best for computations are often unreadable as the final output of the rotational data. The choice of parameters for presentation of such data is dictated not only by clarity but also by existing conventions. Therefore, it is worth to be flexible and have a number of subroutines translating from one parameterization to another.

---

[3] On the other hand, the former approach seems to be more vulnerable to error propagation (e.g. Brown, 1989).

3

Geometry of the Rotation Space

W ITH practically oriented reader in mind, we are going to give a brief account on the structure of the rotation space. Many of the terms used in this chapter are of mathematical origin and have precise definitions in topology, differential geometry, or group theory. This presentation, however, is far from being rigorous. The objective here is an intuitive understanding of some basic issues, and the approach is descriptive without any serious attempts to prove presented statements. For details or deeper understanding the reader is referred to mathematical literature.

As was already explained in chapter 1, proper and improper rotations together constitute a group (i.e., the requirements of closure under composition, associativity, and the existence of the identity and an inverse operation for every element are satisfied). It is isomorphic to the group of orthogonal matrices $O(3)$. Proper rotations alone constitute a group, too. It is isomorphic to the group of special orthogonal matrices $SO(3)$ which is a subgroup of $O(3)$. Actually, $SO(3)$ is an *invariant* subgroup of $O(3)$, which means that for B in $O(3)$ and A in $SO(3)$ the product $B^{-1}AB$ is always in $SO(3)$. $O(3)$ is the direct product of $SO(3)$ and $\{I, -I\}$, i.e. $O(3)$ is isomorphic to the group containing all pairs (A, B) where A is from $SO(3)$ and B is from $\{I, -I\}$ with group product defined by $(A_2, B_2)(A_1, B_1) = (A_2A_1, B_2B_1)$.

A topological space is the least mathematical construct allowing for arguments concerning continuity. Both, $O(3)$ and $SO(3)$ are topological groups, i.e., they are topological spaces and the group operation is a continuous mapping of the group onto itself. The assumed topology is based on the distance of matrices A and B given by $\|A - B\|$, where the (Frobenius) norm $\| \cdot \|$ of a matrix X is defined by $\|X\| := (\text{tr}(X^T X))^{1/2} = (\sum_{ij}(X_{ij})^2)^{1/2}$, i.e., the distance between two matrices is small when their entries are close. Since the determinant of an orthogonal matrix is either $+1$ or -1, the topological space of $O(3)$ consists of two disjoint open subsets (components). On the other hand, there is always a path of proper rotations leading from one orientation to another, which means that $SO(3)$ is *connected*. It is the first of two connected components of $O(3)$. Each element of the second component can be viewed as the product of the inversion and an element of $SO(3)$.

Topological groups[1] belong to the so–called homogeneous spaces. Any point of the space (i.e., element of the group) can be mapped into any other point by composition with an element of the group and that mapping is continuous. Local properties determined for a neighborhood of a given point can be transferred to any other point.

Further on, we will concentrate on the connected component of the orthogonal group, i.e., on $SO(3)$. Because of its relation to 3D-rotations, and thus to the space we live in, $SO(3)$ is one of the most important of all continuous groups. It is also of interest for its relatively rich structure and its relationship to other important groups.

3.1 $SO(3)$ as a Riemannian Manifold

From the geometrical point of view, $SO(3)$ can also be seen as a differential manifold. Generally, a manifold is a space locally resembling the flat (Euclidean) space, but globally it may not conform to it. It is the least structure that supports differentiability and tangents. Formal definition of the N–dimensional manifold is based on the requirement that it can be covered with partially overlapping mutually compatible smooth charts mapping open sets of the manifold onto open sets of R^N. (See, e.g., Spivak, 1965.) Physicists call the charts coordinate systems. For a given chart, each point of the manifold is specified by its coordinates, say x^i. The parameterizations described in chapter 2 provide such charts for $SO(3)$. Since the number of parameters needed in each chart is 3, $SO(3)$ is said to be a manifold of dimension three. For two overlapping charts x^i and x'^i, the coordinates x^i can be expressed through x'^j by using their correspondence to the same point of the manifold, i.e., $x^i = x^i(x'^j)$. By definition of a differential manifold, the two overlapping charts are required to satisfy the condition that the functions $x^i = x^i(x'^j)$ are differentiable on the region of overlap. (The manifold is called analytic if these functions are analytic, i.e., with convergent Taylor series expansion.) This guarantees that differentiation of functions on the manifold makes sense. Also integration on manifolds can be given a meaning. Through the use of charts, a large part of the classical calculus known for R^N can be transformed into the *calculus on manifolds*. This includes calculus of vectors and tensors.

Reasons for investigating vectors on SO(3)
One may think that the esoteric notion of 'a vector tangent to a differential manifold of orientations' is too abstract to be useful. However, the "angular velocity" is nothing else but a vector on $SO(3)$. In theories of deformation of polycrystalline materials "rotation rate fields" are considered. (With deformation by crystallographic slip, it is assumed that grains of similar orientations are rotated with the same rate and a unique rate can be assigned to a given orientation.) The "rotation rate fields" are vector fields on $SO(3)$. For more on this subject see chapter 4. ⊠

The question arises, what can be considered to be a vector (or more generally, a tensor) on a manifold. There are a number of considerably different

---

[1] In the case of topological groups, the mappings ψ defining isomorphism of groups in section 1.5.2 and its inverse ψ^{-1} are required to be continuous.

but equivalent definitions of a vector at a point of a manifold. One of them determines vectors as classes of curves through that point; for a surface in Euclidean space, such class would contain all curves with the same tangent vector and the general case of manifolds is analogous. For our limited purposes, a tensor can be defined as a quantity which satisfies a special transformation rule: at a given point X, the quantity t which in the chart x^i has components $t^{pq...r}{}_{st...u}$ specified by integer indices in the range 1 to N is considered to be a tensor if in the chart x'^i the components of t are given by

$$t'^{ij...k}{}_{lm...n} = t^{pq...r}{}_{st...u} \frac{\partial x'^i}{\partial x^p} \frac{\partial x'^j}{\partial x^q} \cdots \frac{\partial x'^k}{\partial x^r} \frac{\partial x^s}{\partial x'^l} \frac{\partial x^t}{\partial x'^m} \cdots \frac{\partial x^u}{\partial x'^n} \ , \tag{3.1}$$

with the derivatives calculated at the considered point; the upper (lower) indices correspond to contravariant (covariant) transformation principle. A contravariant vector is a particular type of tensor with one index and corresponding transformation rule $t'^i = t^k(\partial x'^i/\partial x^k)$.[2] In the case of Eq.(2.55) used previously, $\partial x'^i/\partial x^p$ and $\partial x^s/\partial x'^l$ are constant and equal to $A^i{}_p$ and $B^s{}_l$, respectively.

It is easy to see the link between the vector transformation rule and classes of curves through a point. With u being a real number in the range from u_0 to u_1, a curve on a manifold is given as the smooth mapping $x^i = x^i(u)$ such that $dx^i/du \neq 0$. With the curve passing through a point given by parameters $p^i = x^i(0)$, the quantity

$$dx^i/du \mid_{u=0} = \lim_{t \to 0} (x^i(t) - p^i)/t \tag{3.2}$$

transforms as $dx'^i/du = (\partial x'^i/\partial x^k)(dx^k/du)$; thus, it is a vector at p^i.

A Riemannian manifold is a differential manifold on which a smooth field of symmetric and positive definite metric tensor g_{ij} is specified. At a given point of the manifold, the metric tensor sends vectors, say v and w, in the tangent space to real numbers via $g_{ij}v^i w^j$, and this can be seen as a generalization of the scalar product of vectors. There may be different metric structures on a manifold with different g fields assigned to it. The coefficients g_{ij}, as components of a covariant tensor, transform between coordinate systems x^i and x'^i via

$$g'_{ij} = g_{mn} \frac{\partial x^m}{\partial x'^j} \frac{\partial x^n}{\partial x'^k} \ . \tag{3.3}$$

In use is also the contravariant form of the metric tensor; its components g^{ij} are defined by $g^{ij} g_{jk} = \delta^i{}_k$.

A curve on a manifold can be parameterized in various ways; it is enough to replace u by another parameter with strictly monotonous dependence on u. The most natural parameterization of a curve is by the curve's arc length s

$$s(u_x) = \int_{u_0}^{u_x} \left(g_{ij}(dx^i/du)(dx^j/du) \right)^{1/2} du \ .$$

---

[2] In modern approach, a contravariant vector is defined as a differential operator acting on scalar fields with scalar results of the action.

Based on this definition, there occurs $(\mathrm{d}s/\mathrm{d}u)^2 = g_{ij}(\mathrm{d}x^i/\mathrm{d}u)(\mathrm{d}x^j/\mathrm{d}u)$. Hence, if the curve is parameterized by arc length (i.e., $x^i = x^i(s)$), then $g_{ij}(\mathrm{d}x^i/\mathrm{d}s)(\mathrm{d}x^j/\mathrm{d}s) = 1$. This is consistent with the expression relating the metric tensor to the 'infinitesimal' distance $\mathrm{d}s$

$$\mathrm{d}s^2 = g_{ij}\mathrm{d}x^i\mathrm{d}x^j .$$

An explicit expression of this type for a given parameterization x^i, is referred to as *line element*. The quantity $\mathrm{d}s$ can be seen as an 'infinitesimal' distance between points differing by $\mathrm{d}x^i$ in their parameters. If finite distances between points of the manifold are known, the corresponding metric tensor can be calculated for a given parameterization by expressing $\mathrm{d}s^2$ through the infinitesimal changes of the parameters. In particular, this applies to the manifold of $SO(3)$ and the rotation parameters. The misorientation angle may serve as a measure of the finite 'distance' between two orientations and only this specific distance will be employed here. The subject will be discussed later but it makes sense to mention here that this distance has the property of being bi–invariant on the $SO(3)$ group: the distance between O and O' is the same as the distance between $O''O$ and $O''O'$, and the same as the distance between OO'' and $O'O''$, where O'' is an arbitrary matrix of $SO(3)$.

Finite distances
For two points (orientations) specified by matrices O and O', the distance in $SO(3)$ is given by $\omega(O,O')$, where

$$\omega(O,O') = \arccos((\mathrm{tr}(O^T O') - 1)/2) ; \tag{3.4}$$

this finite distance was used above and, as will be shown later, it can be considered as natural for $SO(3)$.

In some applications, however, the misorientation angle may be inconvenient as the measure of finite distances between orientations. (Cf. section 5.1.1.) Another distance on $SO(3)$ can be defined by

$$\chi_\bullet^2(O,O') = \|O - O'\|^2/2 = 3 - \mathrm{tr}(O^T O') = 2(1 - \cos(\text{misorientation angle})) , \tag{3.5}$$

where $\|\cdot\|$ denotes the Frobenius norm. Since for ϵ close to zero, $2(1 - \cos(\epsilon)) = \epsilon^2 + \mathcal{O}(4)$, the local metric properties given directly by the misorientation angle are the same as for much simpler $\chi_\bullet(O,O')$.

The distance $\chi_\bullet$ is actually inherited from the linear space of general matrices because the quantity $\|M - M'\|$ is the Cartesian distance between matrices M and M' seen as vectors. In analogous way, the distance inherited from the space of quaternions can be applied. For arbitrary quaternions q and q' it would be $|q - q'|$. In application to orientations, one has to take into account that an orientation is represented by *two* opposite unit quaternions. Therefore, the quaternion based distance is actually defined as

$$\chi_\diamond(q,q') = 2\min\left(|q - q'|, |q + q'|\right) . \tag{3.6}$$

The coefficient 2 was added in order to have the numerical equality $\chi_\diamond \approx \omega$ for close points.

It is easy to check that not only ω but also $\chi_\bullet$ and $\chi_\diamond$ are bi–invariant; e.g. for $\chi_\bullet$,

$$\chi_\bullet(O_L O O_R, O_L O' O_R) = \chi_\bullet(O,O') ,$$

where O_L and O_R are elements of $SO(3)$. Moreover, $\chi_\bullet(O, O') = \chi_\bullet(I, O'O^{-1}) = \chi_\bullet(O'^{-1}, O^{-1})$, and similar relations can be written for $\chi_\diamond$. $\boxtimes$

The definition of a differential manifold provides tools for differentiation of functions on manifolds. One would like to be able to differentiate not only functions but also tensor fields. Let t^i be components of a vector field on the manifold parameterized by x^j. The quantity $\partial t^i/\partial x^j$ does *not* transform as a tensor. Instead, the proper (covariant) derivative of that field is constructed as

$$\partial t^i/\partial x^j + \Gamma^i_{jk}t^k =: t^i_{;j} .$$

With a special rule for transformation between coordinate systems of the (Christoffel) symbols Γ^i_{jk}, the quantities $t^i_{;j}$ constitute a tensor. Assignment of the Christoffel symbols to coordinate systems on the manifold is called a connection. Mathematicians consider manifolds with connection but without metric. However, on a Riemannian manifold, (i.e., manifold with metric) there is a unique canonical connection symmetric in lower indices which is compatible with the metric of the manifold. It is called Levi–Cività connection and is given by

$$\Gamma^i_{jk} = \frac{1}{2}g^{im}\left(\frac{\partial g_{km}}{\partial x^j} + \frac{\partial g_{jm}}{\partial x^k} - \frac{\partial g_{jk}}{\partial x^m}\right) . \tag{3.7}$$

Once a connection is specified, the covariant derivative can be defined for tensors of all ranks in such a way that the operator ';' has properties of the derivative, e.g., the Leibniz identity $(t^j u^k)_{;i} = t^j_{;i}u^k + t^j u^k_{;i}$ is satisfied.

Christoffel symbols transformation rule and covariant derivative of tensors
For coordinate systems x^i and x'^i, the transformation rule of Christoffel symbols is

$$\Gamma'^i_{jk} = \Gamma^l_{mn}\frac{\partial x^m}{\partial x'^j}\frac{\partial x^n}{\partial x'^k}\frac{\partial x'^i}{\partial x^l} + \frac{\partial^2 x^m}{\partial x'^j \partial x'^k}\frac{\partial x'^i}{\partial x^m} .$$

The covariant derivative of a tensor $t^{i...j}_{k...l}$ is given by

$$t^{i...j}_{k...l;m} = \partial t^{i...j}_{k...l}/\partial x^m + \Gamma^i_{mn}t^{n...j}_{k...l} + ... + \Gamma^j_{mn}t^{i...n}_{k...l}$$
$$- \Gamma^n_{km}t^{i...j}_{n...l} - ... - \Gamma^n_{lm}t^{i...j}_{k...n} .$$

The forementioned condition of compatibility between a symmetric connection and the metric is equivalent to the requirement that the covariant derivative of the metric tensor field is equal to zero. It is straightforward to verify this is the case for Christoffel symbols (3.7). $\boxtimes$

Christoffel symbols are essential for establishing parallelism of vectors tangent to different points of a manifold and also for determination of entities corresponding to straight lines in Euclidean spaces. In harmony with intuition, the condition for parallel transport of a vector t^i along a curve $x^j = x^j(s)$ is $t^i_{;j}(dx^j/ds) = 0$. Differently than in the Euclidean geometry, the parallel transport on a curved manifold generally depends on the choice of path. If it does not, it means that the manifold is flat.

Among all paths joining two points of a manifold, we naturally look for a special one which would be the 'shortest'. The 'shortest' paths play the role of straight lines of the Euclidean space. If a parallel transport along $x^j = x^j(s)$ of a vector tangent to that curve at a given point results in vectors tangent to the curve in other points, the curve is called geodesic. More precisely, with $t^i = dx^i/ds$, the curve $x^i = x^i(s)$ is geodesic if the equation $t^i_{;j}t^j = 0$ is satisfied. The explicit form of this equation is

$$\frac{d^2x^i}{ds^2} + \Gamma^i_{jk}\frac{dx^j}{ds}\frac{dx^k}{ds} = 0 . \tag{3.8}$$

Except 'exotic' manifolds, the shortest curve joining two points is geodesic. However, the simple example of the sphere (with great circles as geodesics) shows that not every geodesic segment joining two points is the shortest.

◻ *Geodesically complete manifold*
An example of an exotic manifold mentioned above can be a Euclidean plane with a hole pierced, say, at (0,0); there is no geodesic joining points (1,1) and (-1,-1). A manifold is called geodesically complete if every geodesic can be extended so its parameter s goes to infinity. Every two points of a geodesically complete manifold can be connected by a geodesic. $SO(3)$ is geodesically complete. ⊠

Quantities defined above can be expressed using rotation parameters. If the Rodrigues parameters $\{r^1, r^2, r^3\}$ are taken as coordinates, one chart "almost" covers the whole manifold. The same is true for all other coordinates listed below. With Rodrigues parameters, the metric tensor corresponding to the distance $s = (2\sqrt{c})^{-1} \times$ misorientation angle is given by

$$g_{ij} = (1/c)\left((1 + r^l r^l)\delta_{ij} - r^i r^j\right) / \left(1 + r^k r^k\right)^2 . \tag{3.9}$$

For a reason explained in section 3.4, the constant c is fixed at $c = \pi^{4/3}$. Let us also mention at this point that

$$\sqrt{\det(g)} = 1/(\pi(1 + r^k r^k))^2 . \tag{3.10}$$

Using the contravariant representation of the metric tensor $g^{ij} = c(1 + r^k r^k)\left(\delta_{ij} + r^i r^j\right)$ one obtains the Christoffel symbols

$$\Gamma^i_{jk} = -\frac{\delta_{ij}r^k + \delta_{ik}r^j}{1 + r^l r^l} . \tag{3.11}$$

The above formulas determine the intrinsic geometry of the rotation space in the Rodrigues parameterization. Analogous expressions will be obtained in some other parameterizations.

◻ *Derivation of (3.9)*
This example is given to explain how the components of the metric tensor are calculated. We chose the Rodrigues parameterization because the calculation is relatively brief. Let $(\delta r)^i$ be parameters of the rotation leading from the orientation r^i to a close point $r^i + dr^i$, i.e., the rotation represented by $(\delta r)^i$ is the composition of rotations $(-r^i) \circ (r^k + dr^k)$. Due to Eq.(2.20) $\tan^2(d\omega/2) = (\delta r)^i(\delta r)^i$, and the rotation angle $d\omega$ corresponding to $(\delta r)^i$ is given by $(d\omega)^2 = 4(\delta r)^i(\delta r)^i + \mathcal{O}(((\delta r)^i)^3)$. On the other hand, based on (2.15) the components $(\delta r)^i$ can be expressed as

$$(\delta r)^i = \left(dr^i - \varepsilon_{ijk}r^j dr^k\right) / \left(1 + r^l r^l\right) + \mathcal{O}((dr^i)^2) .$$

Hence,

$$(\mathrm{d}\omega)^2/4 = \left((1 + r^k r^k)\,\mathrm{d}r^i\mathrm{d}r^i - (r^i\mathrm{d}r^i)(r^j\mathrm{d}r^j)\right) / \left(1 + r^l r^l\right)^2 + \mathcal{O}((\mathrm{d}r^i)^3)\ .$$

The distance between two close orientations is determined by $(\mathrm{d}s)^2 = (\mathrm{d}\omega)^2/(4c) = g_{ij}\mathrm{d}r^i\mathrm{d}r^j$; therefore, the components of the metric tensor are given by (3.9). ⊠

▣ *Quaternion coordinates*
For coordinates being the "vector part" $\{q^i, i = 1,2,3\}$ of a quaternion $q = (q^0, q^1, q^2, q^3)$, $(q^0 > 0)$, there occurs

$$g_{ij} = (1/c)\left(\delta_{ij} + q^i q^j/(q^0)^2\right)\ , \quad \text{where} \quad (q^0) := \sqrt{1 - q^i q^i}\ ,$$

$$g^{ij} = c\,(\delta^{ij} - q^i q^j)\ ,$$

$$\sqrt{\det(g)} = \frac{1}{\pi^2 q^0} \tag{3.12}$$

and $\Gamma^i{}_{jk} = c\,q^i g_{jk}$. ⊠

▣ *Modified Rodrigues parameters*
For the modified Rodrigues parameters ζ^i (Eq.2.52), the metric tensor is given by

$$g_{ij} = 4\,\delta_{ij}\,/(c(1 + \zeta^k\zeta^k)^2)\ . \tag{3.13}$$

The square root of the determinant of g has the form

$$\sqrt{\det(g)} = 8/(\pi^2(1 + \zeta^k\zeta^k)^3)\ , \tag{3.14}$$

and the Christoffel symbols are given by $\Gamma^i_{jk} = -2(\delta_{ij}\zeta^k + \delta_{ik}\zeta^j - \delta_{jk}\zeta^i)/(1 + \zeta^k\zeta^k)$. ⊠

▣ *Euler angles*
For Euler angles $\{\varphi_1, \phi, \varphi_2\}$ the components of the metric tensor are

$$g = \frac{1}{4c}\begin{bmatrix} 1 & 0 & \cos\phi \\ 0 & 1 & 0 \\ \cos\phi & 0 & 1 \end{bmatrix}\ ,$$

The tensor can be also written as $g_{ij} = (\delta_{ij} + 2\delta_{1(i}\delta_{j)3}\cos\phi)/(4c)$. Moreover,

$$g^{ij} = \frac{4c}{\sin^2\phi}\left(\delta^{ij} - 2\delta^{1(i}\delta^{j)3}\cos\phi - \delta^{i2}\delta^{j2}\cos^2\phi\right)\ ,$$

$$\sqrt{\det(g)} = \frac{1}{8\pi^2}\sin\phi \tag{3.15}$$

and the Christoffel symbols are

$$\Gamma^i{}_{jk} = \frac{1}{\sin\phi}\left((\delta^{i1}\cos\phi - \delta^{i3})\delta_{1(j}\delta_{k)2} + (\delta^{i3}\cos\phi - \delta^{i1})\delta_{2(k}\delta_{j)3}\right)$$

$$+ \delta^{i2}\delta_{1(j}\delta_{k)3}\sin\phi\ . \tag{3.16}$$

Parentheses () denote symmetrization with respect to indices inside. ⊠

Axis and angle parameters

For completeness, let us add that for the axis/angle coordinates $\{\omega, \vartheta, \psi\}$, where ω is the rotation angle and $\{\vartheta, \psi\}$ are spherical polar coordinates of the rotation axis, g and $\sqrt{\det(g)}$ are given by

$$g = (1/c) \, \mathrm{diag}\left[\frac{1}{4}, \sin^2\frac{\omega}{2}, \left(\sin\vartheta\sin\frac{\omega}{2}\right)^2\right] \,,$$

$$\sqrt{\det(g)} = \frac{1}{2\pi^2}\sin^2\frac{\omega}{2}\sin\vartheta \,, \tag{3.17}$$

and

$$\begin{aligned}
&\Gamma^\vartheta{}_{\vartheta\omega} = \Gamma^\psi{}_{\psi\omega} = \tfrac{1}{2}\cot(\omega/2) \,, && \Gamma^\vartheta{}_{\psi\psi} = -\sin\vartheta\cos\psi \,, \\
&\Gamma^\psi{}_{\vartheta\psi} = \cot\vartheta \,, && \Gamma^\omega{}_{\vartheta\vartheta} = -\sin\omega \,, \\
&\Gamma^\omega{}_{\psi\psi} = -\sin^2\vartheta\sin\omega
\end{aligned}$$

are the only non–zero Christoffel coefficients. ⊠

Isochoric parameters

For the "isochoric" parameters $\rho_i = (3(\omega - \sin\omega)/(4\pi^2))^{1/3} n_i$ (Eq.2.54) the square root of the determinat of g has the trivial form

$$\sqrt{\det(g)} = 1 \,. \tag{3.18}$$

This was already fore–mentioned in section 2.7. ⊠

As the Christoffel symbols were already listed for a number of coordinate systems, and the general equation which the geodesic lines satisfy was specified, the lines' forms in a given system can be determined. With the Rodrigues parameters, the equation of geodesic lines can be written as

$$\frac{\mathrm{d}}{\mathrm{d}s}\left(\frac{1}{1 + r^k r^k}\frac{\mathrm{d}r^i}{\mathrm{d}s}\right) = 0 \,. \tag{3.19}$$

By solving this equation, we find out that a geodesic in Rodrigues space is represented by a straight line. However, it must be kept in mind that some points of the $SO(3)$ manifold are not covered by finite Rodrigues parameters and a second map is needed to include them.

Geodesics in Rodrigues space

Let us define $\xi := 1 + r^k r^k$. After substituting Γ^i_{jk} using Eq.(3.11), the equation of the geodesic line (3.8) takes the form

$$\xi\frac{\mathrm{d}^2 r^i}{\mathrm{d}s^2} - \frac{\mathrm{d}r^i}{\mathrm{d}s}\frac{\mathrm{d}\xi}{\mathrm{d}s} = 0 \,.$$

With $t^i := \mathrm{d}r^i/\mathrm{d}s$, one gets $\xi \mathrm{d}t^i/\mathrm{d}s = t^i \mathrm{d}\xi/\mathrm{d}s$ or $\mathrm{d}(t^i/\xi)/\mathrm{d}s = 0$, i.e., Eq.(3.19). The solution $t^i = a^i \xi$ means that

$$\frac{\mathrm{d}r^i}{\mathrm{d}s} = a^i \xi \,, \tag{3.20}$$

where a^i are integration constants. Multiplication of this equation consecutively by r^i and a^i leads to the system of two first order equations for ξ and $y := a^i r^i$

$$\frac{1}{2}\frac{d\xi}{ds} = y\xi \quad \text{and} \quad \frac{dy}{ds} = a^2\xi , \tag{3.21}$$

where $a^2 = a^i a^i$. After eliminating ξ from the right-hand sides, $a^2 d\xi/ds = dy^2/ds$ and because $a^2\xi - y^2$ is non-negative

$$a^2\xi = y^2 + e^2 , \tag{3.22}$$

where e^2 is another integration constant. Substituting $a^2\xi$ in the second of Eqs. (3.21), we obtain the equation $dy/ds = y^2 + e^2$ with the elementary solution $y = e\tan(es + d)$. Thus, from Eq.(3.22)

$$\xi = \left(\frac{e}{a}\right)^2 \left(1 + \tan^2(es + d)\right) = \left(\frac{e}{a}\right)^2 \frac{1}{\cos^2(es + d)} .$$

Finally, from Eq.(3.20),

$$r^i = a^i \left(\frac{e}{a}\right)^2 \int \frac{ds}{\cos^2(es + d)} = a^i \frac{e}{a^2}\tan(es + d) + b^i .$$

Because of the nature of the geodesic line equation, the parameter s can be replaced by any linear expression of it. Because of this, values for the constants e and d can be suitably chosen. Let $d = 0$. To have the same set $\{r^1, r^2, r^3\}$ corresponding to $s = \omega/(2\sqrt{c})$ and $s = (\omega + 2k\pi)/(2\sqrt{c})$ the quantity e must be a multiple of $\sqrt{c}$; let $e = \sqrt{c}$. Substituting $a^i/(2a^2) \to a^i$, one finally obtains $r^i = a^i \tan(\omega/2) + b^i$. $\boxtimes$

It is obvious that for a fixed rotation axis n and changing rotation angle ω, points $r^i = \tan(\omega/2)n_i$ lie on a straight line through 0. Although not so obvious, it is also true that every straight line in Rodrigues space represents rotations about a fixed axis. The rule of composition of Rodrigues vectors has the property that a rotation applied to the whole set of orientations forming a line in that space transforms that line into another line. Each line which does not contain the origin corresponds to a rotated reference frame and could be mapped back to a line through 0.

Also in quaternion parameterization the equations for a geodesic have relatively simple form

$$\frac{d^2 q^i}{ds^2} + cq^i = 0 . \tag{3.23}$$

As an exercise, one can verify that they are satisfied for q^i being vector components of the quaternion $q = p_c \diamond p$, where p_c is a constant quaternion, and the quaternion p is given by $p^0 = \cos(\omega/2)$ and $p^i = \sin(\omega/2)n_i$ with fixed n_i and $\omega/(2\sqrt{c})$ as the line parameter s. It helps to notice that the general solution of the equation (3.23) has the form

$$q^i(s) = a^i \cos(\sqrt{c}s) + b^i \sin(\sqrt{c}s) ,$$

with a^i and b^i being integration constants, and that the vector part of $p_c \diamond p$ has the same form.

Geodesics and projections
It is easy to check consistency of the analysis of geodesics in Rodrigues and

quaternion spaces. On the sphere of unit quaternions the geodesics are great-circle arcs. Unit quaternions and Rodrigues parameters are related via the gnomic projection; see Eq.(2.47). Generally, the gnomic projection carries great-circle arcs to straight lines.

Moreover, stereographic projection carries circles on a sphere to circles in image space. Hence, in the space of modified Rodrigues parameters ζ_i the geodesic lines are circle arcs. $\boxtimes$

Summarizing, it is clear from the above considerations that geodesic lines in $SO(3)$ are the lines representing rotations about fixed rotation axes. The geodesic can be represented directly by matrices. Let us take the geodesic line from a point represented by the special orthogonal matrix O_1 to one represented by O_2. Let the axis vector and the rotation angle of $(O_1^T O_2)$ be denoted by n and ω $(0 < \omega \le \pi)$, respectively. The points of the geodesic line have the form

$$O(t) = O_1 \Delta(n, t) \,, \tag{3.24}$$

where $\Delta(n, t)$ is the rotation about n by the angle $t = 2\sqrt{c}s$ such that $0 \le t \le \omega$. If $\omega = \pi$ the rotation axis is not unique and the same concerns the geodesic line. Since the parameter t is the rotation angle, we have $\chi_\bullet(O_1, O(t)) = t = \omega - \chi_\bullet(O(t), O_2)$.

Non–convex balls in $SO(3)$
  A subset of a Riemannian manifold is *convex* if the shortest curve between any two points of the subset is unique and lies completely within the subset. It is easy to see that a ball in $SO(3)$ is convex iff its radius is smaller than $\pi/2$. $\boxtimes$

The list of formulas given above provides basic geometrical characteristics of the rotation space in various parameterizations. It is, however, not complete without an important fourth rank tensor called (Riemann) curvature tensor which is a measure of intrinsic curvature.

Curvature tensor
  The formula for that tensor in metric spaces is

$$R^i_{\ jkl} = \partial \Gamma^i_{\ jl}/\partial x^k - \partial \Gamma^i_{\ jk}/\partial x^l + \Gamma^i_{\ ks}\Gamma^s_{\ jl} - \Gamma^i_{\ ls}\Gamma^s_{\ jk} \,.$$

In the case of classical analysis of surfaces in three dimensional Euclidean space, the curvature tensor is related to the fundamental forms of the surfaces. $\boxtimes$

Spaces of constant curvature
  For a point P of a manifold and two linearly independent vectors X and Y given at P, the so–called sectional curvature κ is given by

$$\kappa(P; X, Y) = g_{im} R^i_{\ jkl} X^m Y^j X^k Y^l / \zeta \,,$$

where $\zeta = (g_{ij}X^iX^j)(g_{kl}Y^kY^l) - (g_{ij}X^iY^j)^2$. It can be shown that $\kappa(P; X, Y)$ depends on the 2D vector subspace determined by X and Y but not on the particular choice of vectors representing that subspace. A point is called isotropic if κ does not depend on the choice of that subspace. A Riemannian manifold is said to be of constant curvature if each point of the manifold is isotropic and the sectional curvature has the same value for all points. In the case of classical surfaces in three

dimensional Euclidean space, the sectional curvature is equal to the Gauss curvature.
⊠

With the distance proportional to the misorientation angle, $SO(3)$ is a Riemannian manifold of *constant curvature*. For manifolds of constant curvature κ, the Riemann curvature tensor is simply given by

$$R^i_{\ jkl} = \kappa \left(\delta^i_{\ k} g_{jl} - \delta^i_{\ l} g_{jk} \right) .$$

By calculating this tensor for a particular coordinate system, it can be found out that for $SO(3)$ the constant κ is equal to c.

Manifolds of constant curvature are conformally flat, i.e. the metric tensor g can be expressed as $g_{ij}(P) \propto \delta_{ij}$, with the proportionality coefficient depending on a point of the manifold. This simple form is actually taken by the metric tensor for the modified Rodrigues parameterization; see equation (3.13).

3.2 Exponential Mapping

A special orthogonal matrix can be expressed in a form of an infinite series anlogous to the expansion of the exp function. For a rotation axis given by a unit vector n, let the matrix N be defined by $N_{ij} = -\varepsilon_{ijk} n_k$. From this definition, one has $(N^2)_{ij} = N_{il} N_{lj} = -\delta_{ij} + n_i n_j$ and $N^3 = -N$. Hence, it is easy to see that for $k > 2$

$$N^k = \begin{cases} (-1)^{(k-1)/2} N & \text{for odd } k , \\ (-1)^{(k-2)/2} N^2 & \text{for even } k . \end{cases}$$

With the use of N, the Rodrigues formula (2.21) can be written as

$$O = I + \sin \omega N + (1 - \cos \omega) N^2 . \tag{3.25}$$

Now, let us recall that the trigonometric functions used in this formula have the series representations $\sin \omega = \sum_{k=1(2)}^{\infty} (-1)^{(k-1)/2} \omega^k / k!$ and $1 - \cos \omega = \sum_{k=2(2)}^{\infty} (-1)^{(k-2)/2} \omega^k / k!$. Hence,

$$O = I + N \sum_{k=1(2)}^{\infty} (-1)^{(k-1)/2} \omega^k / k! + N^2 \sum_{k=2(2)}^{\infty} (-1)^{(k-2)/2} \omega^k / k!$$

$$= I + \sum_{k=1(2)}^{\infty} N^k \omega^k / k! + \sum_{k=2(2)}^{\infty} N^k \omega^k / k! = \sum_{k=0}^{\infty} (N\omega)^k / k! .$$

The last expression is analogous to the series expansion of the exponent function. Therefore, the expansion is frequently written as

$$O = \exp(\omega N) . \tag{3.26}$$

The analogy, however, is far from being complete. For example, the relation $\exp(x)\exp(y) = \exp(x + y) = \exp(y)\exp(x)$ which is true for numbers x and y, has no corresponding matrix formula because, generally, matrices do not commute.

The image of a vector v in the rotation by O can be expressed using the series (3.26) $Ov = (\sum_{k=0}^{\infty}(\omega N)^k/k!)v$. The interpretation of particular terms follows from $Nv = n \times v$ and is shown in Fig. 3.1 (Talpe, 1967).

> *Exponential mapping of arbitrary square matrix*
> The series $\exp(X) = \sum_{k=0}^{\infty} X^k/k!$ is convergent for an arbitrary real square matrix X. If matrices X and Y commute, then $\exp(X + Y) = \exp(X)\exp(Y)$. It is easy to see that $\exp(X^T) = (\exp(X))^T$. It can be also shown that $\det(\exp(X)) = \exp(\text{tr}(X))$; thus, $\exp(X)$ is always non–singular. Moreover, $\exp(-X) = (\exp(X))^{-1}$. Obviously, if X is diagonal, so is $\exp(X)$. For non–singular Y, there occurs $\exp(Y^{-1}XY) = Y^{-1}\exp(X)Y$. If X is symmetric, $\exp(X)$ is symmetric and positive definite; if λ is a real eigenvalue of X, then e^{λ} is an eigenvalue of $\exp(X)$.
>
> The matrix $\exp X$ is the limit of the sequence $(I + X/k)^k$ with k growing to infinity
>
> $$\exp X = \lim_{k\to\infty} (I + X/k)^k .$$
>
> The interpretation of this expression in terms of rotations is simple: a finite rotation can be approximated by a composition of a large number k of small steps $I + X/k$.

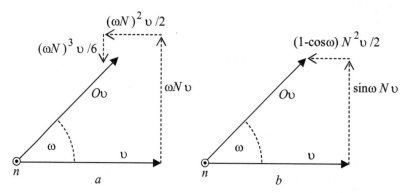

Fig. 3.1. Diagram for the interpretation of the series (3.26) (a). For comparison, the diagram corresponding to Eq.(3.25) is shown in (b).

The matrix $Y(t) = C\exp(tX)$ (where t is a real variable) is a solution to the initial value problem

$$dY/dt = YX \qquad (3.27)$$

for $Y = Y(t)$ such that $Y(0) = C$. Analogously, the matrix $Y(t) = \exp(tX)C$ is a solution to the initial value problem $dY/dt = XY$ with $Y(0) = C$. $\boxtimes$

> *General form of mapping inverse to exp*
> Let the norm of the linear operator A be defined as $\|A\|_L := \sup_v(|Av| / |v|)$, where v is a vector and $|v| := \sqrt{v \cdot v}$. For A such that $\|A - I\|_L < 1$, the series $\log(A) := -\sum_{k=1}^{\infty}(I - A)^k/k$ is convergent and the matrix $X = \log(A)$ satisfies

$\exp(X) = A$. I.e., exp is one–to–one mapping of a small neighborhood of the matrix 0 on the a small neighborhood of I.

What is the region of $SO(3)$ where log is well defined? It is determined by $\|O - I\|_L < 1$. This condition can be written as $\sup_e |Oe - e| < 1$, where e is a unit vector. Since $|Oe - e|^2 = 2(1 - e \cdot Oe)$, the supremum corresponds to the smallest $e \cdot Oe$, i.e., $\sup_e |Oe - e| = \sqrt{2(1 - \cos\omega)} = \sqrt{3 - \text{tr}(O)}$. Hence, the series $\log(O)$ is convergent if O corresponds to a rotation by an angle smaller than $\pi/3$. $\boxtimes$

A special orthogonal matrix was related to antisymmetric matrices in two ways: via exponential mapping $O = \exp(\omega N)$ and by Cayley transformation $O = (I - R)(I + R)^{-1}$. What is the relationship between matrices ωN and R? The simplest way to find out is by using the fact that R can be expressed as $R = -\tan(\omega/2)N$. Hence, $\omega N = -(\omega/\tan(\omega/2))R$. Because of (2.4),

$$\exp\left(-\frac{\omega}{\tan(\omega/2)}(I+O)^{-1}(I-O)\right) = O \qquad \omega \neq 0, \pi$$

where $\omega = \arccos((\text{tr}(O) - 1)/2)$. This is valid also in the limiting case $\omega \to 0$. Thus, log as a mapping inverse to exp, can be defined for the domain beyond $\omega < \pi/3$; for $0 < \omega < \pi$, the log mapping can be expressed in the closed form

$$\log(O) = -\frac{\omega}{\tan(\omega/2)}(I+O)^{-1}(I-O) , \qquad (3.28)$$

and $\log(I) = 0$. Using the identity

$$(\text{tr}(O)I - O)\,O = \det(O)(\text{tr}(O)I - O)^T ,$$

we get $-2(\sin\omega)(I+O)^{-1}(I-O) = \tan(\omega/2)(O - O^T)$, and hence, Eq.(3.28) can be written in a slightly simpler form

$$\log(O) = (\omega/(2\sin\omega))(O - O^T) .$$

Brief calculation shows that the vector associated with the antisymmetric matric $\log(O)$ is the rotation vector ω^i defined by Eq.(2.51). Precisely, for $\sqrt{\omega^i\omega^i} < \pi$ there occurs

$$\omega^i = -\varepsilon_{ijk}(\log(O))_{jk}/2 ;$$

moreover, one has

$$\|\log(O)\|^2 = 2(\omega)^2 . \qquad (3.29)$$

The latter relation follows directly from the previous one.

The exp and log mappings allow the equation (3.24) for the geodesic to be written in a more elegant way. The geodesic line is assumed to go from a point represented by the matrix O_1 to one represented by O_2. Taking into account Eq.(3.24) and the relation $\omega N = \log(O_1^T O_2)$, we can write

$$O(t) = O_1 \exp\left((t/\omega)\log(O_1^T O_2)\right) ; \qquad (3.30)$$

this expression is valid for $0 < \omega < \pi$. It is easy to see that complete geodesics through identity $O(t) = \exp(tN)$ are one parameter subgroups of $SO(3)$.

□ *CBH formula*

For arbitrary matrices X and Y sufficiently close to 0, the matrix $\exp(X)\exp(Y)$ is uniquely determined, and it is given by the so–called Campbell–Baker–Hausdorff (CBH) formula

$$\exp(X)\exp(Y) = \exp(X+Y+[X,Y]/2+([X,[X,Y]]-[Y,[Y,X]])/12+\ldots) \,, \quad (3.31)$$

where $[X,Y]$ denotes the matrix commutator $[X,Y] := XY - YX$. The CBH series formula is valid for a broad class of cases. For $SO(3)$, there exists a corresponding closed form expression. This follows directly from the existence of closed form expressions for log (Eq.3.28) and exp. The latter can be written as:

$$\exp(X) = \begin{cases} I & \text{if } X = 0 \,, \\ I + X(\sin\omega)/\omega + X^2(1-\cos\omega)/\omega^2 & \text{if } X \neq 0 \,, \end{cases}$$

where $\omega = \sqrt{\operatorname{tr}(XX^T)/2}$. See also Engø, 2000. ⊠

□ *Exponential mapping and unit quaternions*

Similarly to matrices, a unit quaternion q can be expressed as

$$q = \exp(v) = \sum_{k=0}^{\infty} (v)^k/k! \,,$$

where v is a quaternion with the scalar part equal to zero, $(v)^0 = e_0$, and $(v)^k := v \diamond v \diamond \ldots \diamond v$ with k factors. An already familiar expression represents the closed form of $q = \exp(v)$: if $|v| \neq 0$ then

$$q^0 = \cos(|v|) \,, \quad q^i = \left(v^i / |v|\right)\sin(|v|) \,,$$

and $q = e_0$ otherwise. The unit quaternion $\exp(v)$ corresponds to the rotation by the angle $2|v|$ about the axis $v_{vect}/|v|$, where v_{vect} is the vector part of v.

In the neighbourhood of e_0 given by $q^0 > 1/2$, the series $\log(q) = -\sum_{k=1}^{\infty}(e_0 - q)^k/k$ is convergent and $\exp(\log(q)) = q$. There exists a closed form expression for $v = \log(q)$ valid for $q^0 > -1$; if $q \neq e_0$ then

$$v^0 = 0 \,, \quad v^i = (q^i/\sqrt{q^j q^j})\arccos(q^0) \,,$$

and $v = 0$ if $q = e_0$. ⊠

3.3 $SO(3)$ as a Lie Group

An analytic manifold which is also a continuous group is referred to as a Lie group if the composition AB of elements A and B of the group is analytic as a function of parameters of A and B. Here, we are concerned only with the so–called linear Lie groups, i.e., continuous *matrix* groups. For linear Lie groups, the tangent space to the manifold at the identity of the group has an additional structure; the space equipped with matrix commutator constitutes a real Lie algebra and there is a natural correspondence between the group elements and vectors of that algebra.

Lie algebra
Formally, a real Lie algebra is a vector space closed under an abstract Lie product $[\cdot,\cdot]_L$ such that for every a, b, c of the space and real α it satisfies: $[a,b]_L = -[b,a]_L$, $[\alpha a + b, c]_L = \alpha[a,c]_L + [b,c]_L$ and the Jacobi identity $[a,[b,c]_L]_L + [b,[c,a]_L]_L + [c,[a,b]_L]_L = 0$. Two Lie algebras with Lie products $[\cdot,\cdot]_1$ and $[\cdot,\cdot]_2$, respectively, are said to be isomorphic if there is a linear one–to–one mapping Ψ from the first algebra to the second such that $[\Psi(a),\Psi(b)]_2 = \Psi([a,b]_1)$. ⊠

Let a vector be interpreted as a class of curves. Let $A = A(t)$ be a curve through identity of the group with $A(0) = I$. A vector of the tangent space at the identity of the group has the form $(dA/dt)|_{t=0}$. In agreement with Eq. 3.27, the matrix $A(t) = \exp(tX)$ is a solution to the initial value problem $dA/dt = XA$ with $A(0) = I$; hence $X = (dA/dt)|_{t=0}$. It can be shown that for curves $A = A(t)$ and $B = B(t)$ such that $A(0) = I$, $B(0) = I$, $X = (dA/dt)|_{t=0}$, $Y = (dB/dt)|_{t=0}$ there occurs $d(A(\sqrt{t})B(\sqrt{t})A(\sqrt{t})^{-1}B(\sqrt{t})^{-1})/dt|_{t=0} = [X,Y]$, i.e., with X and Y in the tangent space, $[X,Y]$ also belongs that space.

Now we will concentrate on $SO(3)$. It was shown in previous section that every special orthogonal matrix can be represented in the form $O = \exp(X)$, where X is an antisymmetric matrix. Since antisymmetric matrices are closed under addition and multiplication by numbers (i.e., for antisymmetric matrices X and Y and real α, the matrix $\alpha X + Y$ is antisymmetric), they constitute a vector space. For 3×3 matrices, the dimension of the space is three, i.e., the same as the dimension of the group. Obviously, the product of antisymmetric matrices does not belong to that space, but the commutator $[X,Y] = XY - YX$ does. The vector space equipped with the additional operation of the commutator satisfies requirements of the (real) Lie algebra.

Based on (3.27), the curve $O(t) = \exp(tX)$ is a solution to the initial value problem $dO/dt = OX$ for $O = O(t)$ such that $O(0) = I$. At $t = 0$, there occurs

$$X = (dO/dt)|_{t=0} \; . \tag{3.32}$$

On the other hand, following Eq.(3.2), $(dO/dt)|_{t=0}$ represents a vector at I. Because of $OO^T = I$, X satisfies $(d(OO^T)/dt)|_{t=0} = X + X^T = 0$, i.e., in agreement with the domain of exp, X must be antisymmetric. Thus, the exponential mapping carries vectors tangent to $SO(3)$ at I to elements of $SO(3)$.

At this point, it is logical to ask about the form of the vector space at points different than I. Let us take a curve $Q = Q(t)$ through $Q(0) = O$. From Eq.(3.2), a vector at O is given by $Y = dQ/dt|_{t=0}$. The curve $P(t) = O^T Q(t)$ satisfies $P(0) = I$ and $dP/dt|_{t=0} = O^T Y$ is antisymmetric. Hence, the vector space tangent to $SO(3)$ at O (not necessarily equal to I) is the space of matrices Y such that the matrices

$$\Omega = O^T Y \tag{3.33}$$

are antisymmetric. The relation $O\Omega = Y$ is invariant under left multiplication by special orthogonal matrices. Proceeding in analogous way, the right invariant form $\Omega' O = Y$ can be obtained. In textures, Ω and Ω' can be seen as measures of angular velocity in the sample and crystal coordinate systems, respectively. In mechanics, they are referred to as velocities in body and spatial reference frames.

Generally, close to the identity, the structure of a Lie group is determined by the structure of its Lie algebra (tangent space at the identity). With vectors E_i constituting a basis of the Lie algebra, any element of the algebra is expressible as a linear combination of E_i. In particular, the commutator of E_i and E_j can be expressed as such combination $[E_i, E_j] = c_{ij}^k E_k$. The coefficients c_{ij}^k – called structure constants – determine the result of the commutator for arbitrary elements of the Lie algebra. The number of independent structure constants is limited by their antisymmetry in lower indices and by the Jacobi identity. Based on their definition, it can be verified that the structure constants are components of a tensor, i.e., with a change of parameterization, they are transformed according to the rule (3.1).

For a Lie group, the Killing form is defined as the bilinear form on its Lie algebra by $(X, Y)_{Killing} = c_{il}^k c_{jk}^l X^i Y^j$, where $X = X^i E_i$ and $Y = Y^i E_i$ are elements of the Lie algebra. For a class of Lie groups to which $SO(3)$ belongs, the Killing form is nondegenerate and negative definite. In these cases, the Killing form times a negative constant factor, say λ, provides the metric tensor of natural choice

$$g_{ij} = \lambda c_{il}^k c_{jk}^l . \tag{3.34}$$

On account of homogeneity, the tensor (3.34) given at the identity can be then transported to all other points by elements of the group. With respect to such metric and with proper choice of λ, the distance of the identity to a point (in the neighborhood of identity) represented by the matrix A is $\sqrt{g_{ij} X^i X^j}$, where the vector X is related to A via $X = \log(A)$. With this metric and constant X, one parameter subgroups of the Lie group are of the form $A(t) = \exp(Xt)$; cf. Eq(3.30).

Now, let us focus again on $SO(3)$. What is the metric (3.34) in this case? Using Eq.(3.29), it is easy to see that the natural Killing form based metric is actually equal (up to a constant factor) to the metric based on the misorientation angle.

In the case of $SO(3)$, the structure of the Lie group is determined by the structure of the Lie algebra almost everywhere. For a suitably chosen parameterization x^i, the basis E_k can be determined with the use of Eq.(3.32) by taking

$$E_k = (\partial O / \partial x_k)\,|_{\text{at the identity}} , \quad \text{where} \quad O = O(x_i) . \tag{3.35}$$

Example
For instance, with the parameters ϕ_1, ϕ_2, ϕ_3 defined in section 2.4.1, the basis obtained in that way has the simple form $(E_i^{(\phi)})_{jk} = \varepsilon_{ijk}$. The parameters ω^i defined by (2.51) lead to the basis $(E_i^{(\omega)})_{jk} = -\varepsilon_{ijk}$. With Rodrigues parameters r^i described in section 2.2 the basis is $(E_i^{(r)})_{jk} = -2\varepsilon_{ijk}$. However, not all coordinates are suitable for determining basis vectors in that way. The Euler angles $\varphi_1, \phi, \varphi_2$ give the vectors

$(E_1)_{jk} = -\varepsilon_{1jk}$, $(E_2)_{jk} = -\varepsilon_{2jk}$ and $(E_3)_{jk} = -\varepsilon_{1jk}$, respectively; these vectors are not linearly independent and cannot serve as a basis. $\boxtimes$

Example for vector transformation rule

This is a good point for an exercise. Let us consider a vector at the identity with components t^i and t'^i given in the bases $E_i^{(\omega)}$ and $E_i^{(r)}$, respectively, i.e., $t^i E_i^{(\omega)} = t'^i E_i^{(r)}$. Since the bases differ by the factor of 2, one has $t^i = 2t'^i$. The question arises, whether this is consistent with the formula (3.1) $t'^i = t^j(\partial r^i/\partial \omega^j)$ for transformation of vector components. In order to check that, r^i must be expressed as a function of ω^j; that relation is given by $r^i(\omega^j) = (\omega^i/\sqrt{\omega^k\omega^k})\tan(\sqrt{\omega^l\omega^l}/2)$. Now, brief calculation shows that the limit of the derivative $\partial r^i/\partial \omega^j$ for $\sqrt{\omega^k\omega^k}$ approaching zero is $\delta_{ij}/2$. This means that $t'^i = t^j\delta_{ij}/2 = t^i/2$, which is in agreement with the relation between bases at the identity. $\boxtimes$

Metric in Rodrigues parameterization obtained from (3.34)

With a metric given at one point of $SO(3)$, the complete metric field can be determined due to the space homogeneity. In the Rodrigues parameterization, with the basis $(E_i^{(r)})_{jk} = -2\varepsilon_{ijk}$, the structure constants are $c^i_{jk} = 2\varepsilon_{ijk}$, and the metric tensor at the identity ($r^i = 0$) calculated from (3.34) is $g_{ij}(0) = -8\lambda\delta_{ij}$. The neighborhood of the identity is mapped onto a neighborhood of other points by composition with a given rotation. Let p^i be parameters of a fixed point p in the Rodrigues space, and let r^i be variable parameters. The infinitesimal distance must be the same in both neighborhoods, i.e.,

$$g_{ij}(0)\,\mathrm{d}r^i\mathrm{d}r^j = g_{ij}(p)\,\mathrm{d}(p \circ r)^i\mathrm{d}(p \circ r)^j \;, \tag{3.36}$$

where $g_{ij}(p)$ are entries of the metric tensor at p. Based on (2.15), the factor $\mathrm{d}(p \circ r)^i$ at $r^i = 0$ is

$$\mathrm{d}(p \circ r)^i\,|_{r^i=0} = (\delta_{ik} + \varepsilon_{iku}p^u + p^i p^k)\,\mathrm{d}r^k \;.$$

Hence, Eq.(3.36) takes the form

$$-8\lambda\delta_{ij}\,\mathrm{d}r^i\mathrm{d}r^j = g_{ij}(p)(\delta_{ik} + \varepsilon_{iku}p^u + p^i p^k)(\delta_{jl} + \varepsilon_{jlw}p^w + p^j p^l)\,\mathrm{d}r^k\mathrm{d}r^l \;,$$

and this must be true for all $\mathrm{d}r^i$. Therefore,

$$-8\lambda\delta_{kl} = g_{ij}(p)(\delta_{ik} + \varepsilon_{iku}p^u + p^i p^k)(\delta_{jl} + \varepsilon_{jlw}p^w + p^j p^l) \;.$$

Since the inverse to $(\delta_{ik} + \varepsilon_{iku}p^u + p^i p^k)$ is given by $(\delta_{kj} - \varepsilon_{kjs}p^s)/(1 + p^m p^m)$, the metric tensor at p is given by

$$g_{ij}(p) = -8\lambda\delta_{kl}(\delta_{ki} - \varepsilon_{kis}p^s)(\delta_{lj} - \varepsilon_{ljt}p^t)/(1 + p^m p^m)^2$$
$$= -8\lambda((1 + p^k p^k)\delta_{ij} - p^i p^j)/(1 + p^m p^m)^2 \;.$$

For $\lambda = -1/(8c)$, this is the same expression as (3.9). $\boxtimes$

It is proved in the Lie group theory that every real Lie algebra is isomorphic to the real Lie algebra of some linear Lie group. But is this group unique? Does the structure of a Lie algebra determine the structure of the Lie group globally? The answer to these questions is 'no'. Based on this answer, one may ask about other Lie groups which have the same Lie algebra as $SO(3)$.

This question can be answered directly but it is worth to say a bit more on the subject.

The story is based on the topology of the group. A path on a topological space is a continuous mapping of a segment, say $[0, 1]$, into that space. A path, say P, is said to be closed if $P(0) = P(1)$. A path can be continuously deformed to get other paths. Some closed paths can be shrunk to a point by such deformation. There is a special class of connected topological spaces called simply connected topological spaces which can be roughly described as spaces "without holes". More precisely, a connected topological space is simply connected if every closed path can be shrunk to a point. It is intuitively obvious that the sphere S^2 is simply connected and the torus $S^1 \times S^1$ is not.

Also, a Lie group is called simply connected if it is simply connected as a topological space. One of the important results of the Lie group theory is that for each Lie group G there exists a simply connected Lie group $\overline{G}$ such that G and $\overline{G}$ are *locally* isomorphic.[3] The simply connected group $\overline{G}$ is called the *covering group* of G.[4] The covering group is unique up to an isomorphism. Moreover, the groups G and $\overline{G}$ have isomorphic real Lie algebras.

$SO(3)$ is *not* simply connected. In order to see that, it is enough to find a path which is not contractible to a point. In the parameterization by the rotation vector described in section 2.7, the rotations by the angle of π are represented by two antipodal points on the surface of the parametric ball. Now, take the path corresponding to the full 360 degrees rotation about a fixed axis. Let the path start at the center of the parametric ball. With growing path parameter, the path reaches the ball surface and at the same time appears at the antipodal point, and then approaches its ending point at the center (Fig. 3.2). There is no way to shrink this path to a point because continuous deformations of the path cannot bring together the antipodal points. A closed path which does not reach the surface is contractible to a point. Also a closed path which crosses the surface twice can be shrunk to a point. Therefore there are only two classes of paths on $SO(3)$; it is said to be doubly connected.

It is proved in topology that for $N \geq 3$ the groups $Spin(N)$ described in section 1.5.2 are simply connected. $Spin(3)$ is isomorphic to the group of unit quaternions and to $SU(2)$. Moreover, the neighborhood of the identity in $SO(3)$ has the same group structure as the neighborhood of the identity in $SU(2)$. Thus, $SU(2)$ is the universal covering group of $SO(3)$. There is a two–to–one homomorphism of $SU(2)$ onto $SO(3)$; the homomorphism is continuous and locally one–to–one.[5] Finally, the real Lie algebra of $SU(2)$ is isomorphic

---

[3] I.e., there exist a diffeomorphic (differentiable and with differentiable inverse) mapping, say ψ, between neighborhoods of identities such that $\psi(AB) = \psi(A)\psi(B)$ whenever A, B and AB are in some neighborhood of the identity of the first group, and $\psi(A)$, $\psi(B)$ and $\psi(AB)$ are in some neighborhood of the identity of the second group.

[4] G is isomorphic to the factor group $\overline{G}/K$, where K is a discrete subgroup of the center of $\overline{G}$. The center of a given group $\overline{G}$ is a subgroup of $\overline{G}$ consisting of all elements which commute with every element of $\overline{G}$.

[5] The quaternions $(1, 0, 0, 0)$ and $(-1, 0, 0, 0)$ constitute the center of the group of unit quaternions. Analogously, the the matrices $\pm I_2$ constitute the center, say Z, of $SU(2)$ and there occurs $SO(3) \simeq SU(2)/Z$.

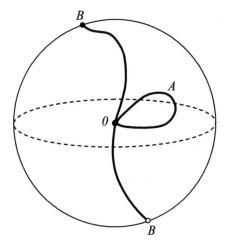

Fig. 3.2. Two closed paths in the parametric space of rotations. Path B cannot be shrunk to a point.

to the algebra of antisymmetric 3×3 matrices with matrix commutator as the algebraic operation, i.e., it is isomorphic to the Lie algebra of $SO(3)$.

> *Universal covering group of $SO(2)$*
> The universal covering group of $SO(2)$ is the group of real numbers with addition as the group operation. To see that intuitively, it is enough to identify elements of $SO(2)$ with points of a circle and real numbers with points of a line, and to wrap the line around the circle. This gives infinite–to–one homomorphism of the groups and their local isomorphism. ⊠

Let us close this section with a description of two other properties of the group of special orthogonal matrices. As any Lie group, $SO(3)$ is orientable. It means that there exists a covering such that for any two overlapping charts x^i and x'^k, one has $\det[\partial x^i / \partial x'^j] > 0$ in the region where they overlap. In relation to that, $SO(3)$ as a manifold is described as parallelizable. An N dimensional manifold is said to be parallelizable if it admits N linearly independent smooth vector fields. (Generally, every Lie group is parallelizable.)

$SO(3)$ is also characterized as *compact*. This adjective has far reaching consequences for Lie groups. For subsets of the Euclidean space such characterization is equivalent to the statement that a set is closed and bounded. The issue is more complex in the case of manifolds but, in practice, for linear Lie groups, it means that there exist parameters contained in closed finite intervals and covering the manifold. Generally, compact manifolds can be covered by finitely many charts. Moreover, continuous real functions on compact manifolds are bounded.

3.4 Integration on $SO(3)$

Frequently, while dealing with functions determined on the orientation space, one encounters the problem of integrating over that space or over a part of it. A proper setting for integration on manifolds is the calculus of differential forms. For our limited purposes it is sufficient to stay within a more elementary approach. An integral on a manifold makes sense only if it is independent of coordinates (parameterization) chosen for calculations. This is also true for integration in R^n. Therefore, this section begins with recalling the rule for the change of variables in elementary integral calculus in R^n.

Assume a function f given in R^n depends explicitly on variables x^i. The function is integrated over certain region, say V_x,

$$\int_{V_x} f(x^i)\, dx^1\, dx^2...dx^n \quad .$$

If the variables are changed to x'^k with $x^i = x^i(x'^j)$, then the above integral can be expressed as

$$\int_{V_{x'}} f(x^i(x'^j))\, |\det J|\ dx'^1\, dx'^2...dx'^n \quad .$$

where $V_{x'}$ is the image of the region V_x in the transformation $x^i = x^i(x'^j)$ and the Jacobian matrix J is defined by $J^i_j = \partial x^i/\partial x'^j$.

Now, let f have a new meaning; it is a function on the domain V on an orientable manifold. A point p of that manifold can be specified by x^i or x'^j. Coordinates x^i and x'^j represent two different charts on the manifold. The transformation of the metric tensor caused by the change of the coordinate system is given by (3.3). Hence, $\sqrt{\det g'} =|\det J|\ \sqrt{\det g}$. The integral of f over V does not depend on parameterization if it is calculated according to

$$\int_{V_x} f(p(x))\sqrt{\det g}\ dx^1\, dx^2...dx^n = \int_{V_{x'}} f(p(x'))\sqrt{\det g'}\ dx'^1\, dx'^2...dx'^n$$

where V_x and $V_{x'}$ are images of V in the corresponding charts. Sometimes the above relation is written as the equality of volume elements $\sqrt{\det g}\, dx^1\, dx^2...dx^n = \sqrt{\det g'}\, dx'^1\, dx'^2...dx'^n$.

On the other hand, with the group structure involved, another requirement appears. For a function given on a finite group, there occurs $\sum_A f(A) = \sum_A f(AA') = \sum_A f(A'A)$. This can be generalized to connected Lie groups with summation replaced by integration over parameters

$$\int f(A)w(A)\mathrm{d}A = \int f(AA')w(AA')\mathrm{d}A = \int f(A'A)w(A'A)\mathrm{d}A$$

where $\mathrm{d}A$ stands for $dx^1\, dx^2...dx^n$, w is a weight function, and the integration is over the whole group. But will such a relation exist for each Lie group? The answer is no. For compact Lie groups, however, the invariant integral always exists and is finite for continuous functions; moreover, the weight function is unique up to a constant factor. With invariant metric g, there occurs

$w \propto \sqrt{\det g}$. The quantity $\sqrt{\det g}\, dx^1\, dx^2 \ldots dx^n$ is referred to as the *invariant volume element*. It is natural to fix the constant factor by assuming that the integral of $f(A) = 1$ over the whole group is equal to 1.

We have already mentioned that $SO(3)$ is compact, so all of the above applies to it. Thus, to perform integration on the rotation space in a given parameterization, one needs the square root of the determinant of the metric tensor. That is why proper expressions were listed in Eqs. (3.10, 3.12, 3.14 3.15, 3.17, 3.18). The value of the coefficient c used in the previous section is set in such a way that the integral of $f(O) = 1$ over the whole group equals 1.

It is not always possible to cover the integration domain with one chart. General solution to that practical problem is quite complicated and involves the so–called partition of unity. In many cases, however, one chart may cover the complete domain of integration except "a set of measure zero", i.e., a set to small to contribute to integrals of smooth functions. In particular, such charts exist for $SO(3)$.

Volume of a ball

As an exercise, the volume of a ball of radius ω_0 in the rotation space will be calculated. This can be easily done using the axis–angle coordinates $\{\omega, \vartheta, \psi\}$. Because of the homogeneity of the space, the center of the ball can be positioned at the identity; in this case, the radius is measured by the parameter ω. Based on (3.17), the volume $V(\omega_0)$ is given by

$$V(\omega_0) = \int_0^{2\pi} d\psi \int_0^{\pi} d\vartheta \int_0^{\omega_0} d\omega \left(\frac{1}{2\pi^2} \sin^2(\omega/2) \sin\vartheta \right) = \frac{1}{\pi}(\omega_0 - \sin\omega_0) \ .$$

This expression also follows directly from the formula for the volume of a ball in Euclidean space $V(\omega_0) = (4/3)\pi(\rho)^3$, where $\rho = \rho(\omega_0)$ is the 'radius' $\rho = \sqrt{\rho^i \rho^i}$ in the isochoric parameter 'space'; cf. Eq.(2.54).

The ball of radius $\omega_0 = \pi$ covers the whole space and has the volume equal to $V(\pi) = 1$. Rotations by angles not larger than $\pi/2$ constitute only $V(\pi/2) = 1/2 - 1/\pi \approx 0.18169$ of the whole space. $\boxtimes$

4

More on Small Orientation Changes

INFINITESIMALLY small changes of orientations are essential for rotational kinematics. They are important also in textures; especially for the analysis of texture evolution in plastic deformation of polycrystals.

4.1 Vector of Infinitesimal Rotation

Orientation changes linear in orientation parameters are usually described by a certain vector, which will be denoted here by ϖ. Let an orientation be represented by an orthogonal matrix O. By adding small quantities grouped in the matrix $\mathrm{d}O$ to the entries of O, we get the matrix $O + \mathrm{d}O$ which is orthogonal if $\mathrm{d}O$ satisfies

$$O^T \mathrm{d}O + \mathrm{d}O^T O + \mathrm{d}O^T \mathrm{d}O = 0 \ . \tag{4.1}$$

The misorientation δO between $O + \mathrm{d}O$ and O can be written in the form

$$\delta O = O^T (O + \mathrm{d}O) = I + \Omega \ , \quad \text{where} \quad \Omega = O^T \mathrm{d}O \ . \tag{4.2}$$

For small entries of $\mathrm{d}O$, with the quadratic term $\mathrm{d}O^T \mathrm{d}O$ in Eq.(4.1) neglected, the matrix Ω is antisymmetric, i.e., $\Omega + \Omega^T = 0$. The relationship between Ω and tangent vectors is embodied in Eq.(3.33). At the identity, the matrix Ω is directly related to the matrices given by (3.35) $\Omega \mid_{\text{at the identity}} = E_k \mathrm{d}\xi^k$, where ξ^k are the rotation parameters. With antisymmetric Ω, the matrix δO can be expressed as

$$\delta O_{ij} = \delta_{ij} + \varepsilon_{ijk} \varpi_k \ ,$$

where ϖ is the vector of infinitesimal rotation given by $\varpi_i = \varepsilon_{ijk} \Omega_{jk}/2$.

The vector ϖ_i depends linearly on infinitesimal changes of other orientation parameters. For the Rodrigues parameters r^i, based on the relation (2.14), $\mathrm{d}O$ is given by

$$\mathrm{d}O_{ij} = (2/(1 + r^k r^k)^2) \left((1 + r^m r^m)(r^i \mathrm{d}r^j + r^j \mathrm{d}r^i) - 2(\delta_{ij} + r^i r^j) r^m \mathrm{d}r^m + \varepsilon_{ijn}(2r^n r^m \mathrm{d}r^m - \mathrm{d}r^n (1 + r^m r^m)) \right) \ .$$

Hence, the definition of the vector ϖ leads to

$$\varpi_i = \frac{-2}{(1 + r^m r^m)} \left(\delta_{ik} - \varepsilon_{ijk} r^j \right) \mathrm{d}r^k \ . \tag{4.3}$$

The inverse relation has the form

$$\mathrm{d}r^i = -\frac{1}{2} \left(\delta_{ik} + r^i r^k + \varepsilon_{ijk} r^j \right) \varpi_k \ . \tag{4.4}$$

Alternative derivation
    The same can be obtained by calculating components δr^i of the Gibbs vector $\delta r = (-r) \circ (r + dr)$

$$\delta r^i = \frac{\delta_{ij} + \varepsilon_{ijk} r^k}{1 + r^l r^l} \mathrm{d}r^j \ .$$

The orthogonal matrix δO and the corresponding Rodrigues vector $\delta \mathbf{r}$ are related via Eq.(2.14)

$$\delta O_{ij} = \delta_{ij} - 2\varepsilon_{ijk} \frac{\delta_{kl} + \varepsilon_{klm} r^m}{1 + r^n r^n} \mathrm{d}r^l \ .$$

By comparing this with $\delta O_{ij} = \delta_{ij} + \varepsilon_{ijk} \varpi_k$, Eq.(4.3) is obtained. $\boxtimes$

Proceeding in an analogous way, the following relations between the vector of infinitesimal rotation and the increment of quaternion coordinates q^i, $i = 1, 2, 3$ can be derived:

$$\varpi_i = \frac{-2}{q^0} \left((q^0)^2 \delta_{ij} + q^i q^j + \varepsilon_{ijk} q^k q^0 \right) \mathrm{d}q^j \quad \text{and} \quad \mathrm{d}q^i = -\frac{1}{2} \left(q^0 \delta_{ij} - \varepsilon_{ijk} q^k \right) \varpi_j \ .$$

The first relation means that $-\varpi_i/2$ is the vector part of the quaternion $q^* \diamond \mathrm{d}q$ with $\mathrm{d}q^0$ determined by $q^\mu \mathrm{d}q^\mu = \mathrm{d}(q^\mu q^\mu)/2 = 0$; the scalar part of $q^* \diamond \mathrm{d}q$ is zero.
    Based on the relation between axis/angle parameters (n, ω) and the entries of the orthogonal matrix (Eq.2.21), we have

$$\varpi_i = - \left(n_i \mathrm{d}\omega + (\sin \omega \delta_{ik} - (1 - \cos \omega)\varepsilon_{ijk} n_j)\mathrm{d}n_k \right) \ ,$$

where $n_k \mathrm{d}n_k = \mathrm{d}(n_k n_k)/2 = 0$. It is obvious here that in the 'axis–angle' parameterization, a change of the n_k parameters corresponds to a rotation; this is not always clear, when infinitesimal rotations are defined via $n_i \mathrm{d}\omega$ with fixed rotation axis.
    The vector ϖ and increments of Euler angles are related by

$$\begin{aligned}
\varpi_1 &= \quad \cos \varphi_1 \, \mathrm{d}\phi + \sin \varphi_1 \sin \phi \, \mathrm{d}\varphi_2 \\
\varpi_2 &= \quad \sin \varphi_1 \, \mathrm{d}\phi - \cos \varphi_1 \sin \phi \, \mathrm{d}\varphi_2 \\
\varpi_3 &= \mathrm{d}\varphi_1 \qquad\qquad + \qquad \cos \phi \, \mathrm{d}\varphi_2
\end{aligned} \tag{4.5}$$

and

$$\begin{aligned}
\mathrm{d}\varphi_1 &= - \cot \phi \sin \varphi_1 \, \varpi_1 + \cot \phi \cos \varphi_1 \, \varpi_2 + \varpi_3 \\
\mathrm{d}\phi &= \quad\quad \cos \varphi_1 \, \varpi_1 + \quad\quad \sin \varphi_1 \, \varpi_2 \\
\mathrm{d}\varphi_2 &= \quad \csc \phi \sin \varphi_1 \, \varpi_1 - \csc \phi \cos \varphi_1 \, \varpi_2
\end{aligned} \ .$$

□ *Vector of infinitesimal rotation in crystal frame*
Analogously to (4.2), one can consider $\delta O' = (O + \mathrm{d}O)O^T = I + \Omega'$, where
$\Omega' = \mathrm{d}OO^T$. As was already mentioned in chapter 3, the difference between Ω
and Ω' is associated with the choice of coordinate system in which the change of
orientation is given. In the convention used in textures, Eq.(4.2) corresponds to the
orientation change given in the sample coordinate system while $\delta O'$ is the change
in the crystal coordinate system.

The relations of $\varpi'_i = \varepsilon_{ijk}\Omega'_{jk}/2$ and parameter increments can be derived in the
same way as in the case of ϖ. Brief calculation shows that for Rodrigues parameters

$$\varpi'_i = \frac{-2}{(1 + r^k r^k)}\left(\delta_{ik} + \varepsilon_{ijk}r^j\right)\mathrm{d}r^k .$$

The relation of the vector ϖ' to the increments of Euler angles has the form

$$\begin{aligned}
\varpi'_1 &= \sin\varphi_2 \sin\phi\, \mathrm{d}\varphi_1 + \cos\varphi_2\, \mathrm{d}\phi \\
\varpi'_2 &= \cos\varphi_2 \sin\phi\, \mathrm{d}\varphi_1 - \sin\varphi_2\, \mathrm{d}\phi \\
\varpi'_3 &= \qquad \cos\phi\, \mathrm{d}\varphi_1 \qquad\qquad + \mathrm{d}\varphi_2 .
\end{aligned}$$

The long direct derivation of the above formulas (from $\Omega' = \mathrm{d}OO^T$) can be avoided
by replacing the variables $\varphi_1, \phi, \varphi_2$ and ϖ in Eq.(4.5) by $-\varphi_2, -\phi, -\varphi_1$ and $-\varpi'$,
respectively. The set of inverse relations

$$\begin{aligned}
\mathrm{d}\varphi_1 &= \quad \csc\phi \sin\varphi_2\, \varpi'_1 + \csc\phi \cos\varphi_2\, \varpi'_2 \\
\mathrm{d}\phi &= \qquad \cos\varphi_2\, \varpi'_1 - \qquad \sin\varphi_2\, \varpi'_2 \\
\mathrm{d}\varphi_2 &= -\cot\phi \sin\varphi_2\, \varpi'_1 - \cot\phi \cos\varphi_2\, \varpi'_2 + \varpi'_3 .
\end{aligned}$$

is well known in crystallographic textures (Bunge, 1970). It is applied to analysis of
crystallite rotation rates in plastically deformed polycrystals. ⊠

With $\chi_\bullet$ defined by (3.5), the quantity $\chi_\bullet(\delta O, I)$ is a scalar measure of an
infinitesimal orientation change. We have

$$\chi^2_\bullet(\delta O, I) = \|\Omega\|^2/2 = \varpi_i \varpi_i .$$

The measure can be expressed via coordinates. Take the Rodrigues parameters. Brief calculation based on Eqs. (3.9) and (4.4) shows that

$$\varpi_i \varpi_i = (4\pi^{4/3})g_{ij}\mathrm{d}r^i\mathrm{d}r^j . \tag{4.6}$$

This also follows directly from the relationship between $\chi_\bullet$ and the metric.
Relations analogous to (4.6) hold for other parameterizations.

The components ϖ_i are *not* differentials. This means that there are no
functions ψ_i such that $\mathrm{d}\psi_i = \varpi_i$. If such ψ_i had existed, second partial derivatives would have been symmetric and that does not occur[1]. In consequence,
if t is a parameter along a continuous curve in orientation space, the integrals
$\int_{t_0}^{t_1} \varpi_i(t)\mathrm{d}t$ depend not only on the starting and final orientations, but also
on the rotation path.

---

[1] E.g., in Rodrigues parameterization, $\partial\psi_i/\partial r^k = -2(\delta_{ik} - \varepsilon_{ijk}r^j)/(1 + r^m r^m)$ and
$\partial\psi_i/\partial r^k \partial r^l \neq \partial\psi_i/\partial r^l \partial r^k$ if $k \neq l$.

4.2 Rotation Rate Field and Continuity Equation

The rotation rate field $\varpi = \varpi(O)$ and orientation flow field $J = J(O)$ have been used to model and investigate the evolution of crystallographic textures during plastic deformation of polycrystals. The fields are related by $J(O) = f(O)\varpi(O)$, where f is the orientation distribution. Knowing (from one of the crystal deformation theories) an approximation of the lattice rotation rates in the crystal coordinate system, one can get the rates in the sample coordinate system.

With continuous rotations of particular crystallites, the orientations flow continuously in the orientation space. Accordingly, the orientation distribution evolves with the assumed "time" t, i.e., it is a function of O and t : $f = f(O,t)$. Also the rotation rate is generally time–dependent. Since crystallite orientations are neither created nor destroyed, for sufficiently smooth f and J, the principle of conservation of orientations can be formally expressed in the form of the continuity equation (Clément, 1982)

$$\frac{\partial f}{\partial t} + \mathrm{div}(J) = 0 \ .$$

In particular coordinates, the divergence operator $\mathrm{div}(J) = J^i{}_{;i}$ is calculated using Christoffel symbols listed in chapter 3.

For example
In the Rodrigues parameterization, from Eq.(3.11), the divergence of J is given by $\mathrm{div}(J) = \partial J^i/\partial r^i + \Gamma^i_{ij}J^j = \partial J^i/\partial r^i - 4J^i r^i/(1 + r^k r^k)$. ⊠

The derivation of the continuity equation is analogous to that in electro– or fluid dynamics and is based on the Gauss' theorem generalized to orientable compact manifolds. With the orientation distribution f normalized over the orientation space $(\int_{SO(3)} f\mathrm{d}O = 1)$, there occurs

$$\int_{SO(3)} \mathrm{div}(J)\mathrm{d}O = - \int_{SO(3)} \frac{\partial f}{\partial t}\mathrm{d}O = -\frac{\mathrm{d}}{\mathrm{d}t} \int_{SO(3)} f\mathrm{d}O = 0 \ ,$$

i.e., the integral of the divergence of the flow field vanishes. (It s shown in differential geometry that this is generally true for vector fields of class C_1 on compact orientable manifolds.)

If the rotation rate field is *constant* in time, i.e., $\varpi = \varpi(O)$, the texture at t is given by

$$f(O,t) = \exp(t\mathcal{D}_\varpi)\,f(O,0) \ , \tag{4.7}$$

where $\mathcal{D}_\varpi f = -\mathrm{div}(f\varpi)$, and $\exp\alpha$ is interpreted as $\sum_{k=0}^{\infty} \alpha^k/k!$. At an orientation O_0 at which the rotation rate vanishes, the dependence of the texture function on t is exponential. At such a point $(\mathcal{D}_\varpi f)(O_0,0) = f(O_0,0)\mathrm{div}(\varpi)(O_0)$, and hence $f(O_0,t) = f(O_0,0)\exp(-t\mathrm{div}(\varpi))(O_0)$.

Outline of the derivation of (4.7)
For sufficiently smooth f, the function can be expressed in the form of the McLaurin series with respect to (finite) t

$$f(O,t) = \sum_{k=0}^{\infty} f_k(O)t^k \ . \tag{4.8}$$

After substituting f in the continuity equation, we get $\sum_{k=1}^{\infty} (kf_k + \mathrm{div}(f_{k-1}\varpi)) \, t^{k-1} = 0$. This is a recurrence formula for f_k

$$f_k = -(1/k)\mathrm{div}(f_{k-1}\varpi) = (1/k)\mathcal{D}_\varpi(f_{k-1}) = (1/k!)\mathcal{D}_\varpi^k(f_0) \ .$$

Hence, the solution (4.7) is obtained based on Eq.(4.8). ⊠

4.3 Short Excursion into Mechanics

The notion of angular velocity familiar from elementary kinematics, is directly related to the vector of infinitesimal rotation. Motion in non–inertial reference frames and rigid body motion are flag applications of the rotation rate fro-malism. The material below is given to recap briefly some basic aspects of mechanics in our notation; this may be helpful for better understanding of the rotation rate issue.

4.3.1 Motion of a particle in a rotating coordinate system

Let x' be a vector determining a location of a particle with respect to an inertial reference system. (All vectors given in the inertial system will be distinguished by the prime.) Transformation to a non–inertial system rotating about the common origin of the systems has the form $x' = Ox$, where O is an orthogonal matrix. With the dot denoting the time derivative,

$$\dot{x}' = \dot{O}x + O\dot{x} = O(\dot{x} + \Omega x) \ , \tag{4.9}$$

where $\Omega = O^T\dot{O}$ represents the instant orientation change per time unit, i.e., the rotation rate. Roughly, the velocity in the inertial system is the sum of the velocity in the non–inertial system and the velocity appearing due to the system rotation. Using the "total" time derivative D_t with connection Ω,

$$D_t = \frac{\mathrm{d}}{\mathrm{d}t} + \Omega \ ,$$

Eq.(4.9) takes a simpler form

$$\dot{x}' = OD_t x \ .$$

The operator D_t can be expressed as $D_t = O^T(\mathrm{d}/\mathrm{d}t)O$. This expression is convenient for transforming particle dynamics equations from the inertial sys-tem to rotating systems; next subseciton shows that this is done simply by replacing $\mathrm{d}/\mathrm{d}t$ with D_t.

4.3.2 Newton's equations in a rotating coordinate system

In the inertial coordinate system, the equations of motion of a particle of mass m is $\dot{p}' = F'$, where p' is the vector of momentum $p' = m\dot{x}'$, and $F' = F'(x', \dot{x}')$ is a force acting on the particle. With the momentum in the rotating coordinate system defined by $p = mD_t x$, one has $Op = mOD_t x = m\dot{x}'$ and thus, $p' = Op$. The frame invariant form of the Newton's equations is

$$D_t p = F ,$$

with F given by $F' = OF$. It follows from $D_t p = O^T (\mathrm{d}/\mathrm{d}t) O p = O^T (\mathrm{d}/\mathrm{d}t) p' = O^T F' = F$. Thus, the equations of motion in the rotating coordinate system are analogous to the inertial case with each vector, say y', replaced by $O^T y'$ and time derivative replaced by the "total" derivative.

The left side of $D_t p = F$ can be expressed as $D_t p = m D_t^2 x = m D_t (\dot{x} + \Omega x) = m\ddot{x} + m\dot{\Omega}x + 2m\Omega\dot{x} + m\Omega\Omega x$, which is a form of the familiar expression with the third and fourth terms representing Coriolis and centrifugal forces, respectively.

The equations for angular momentum $l' = x' \times p'$ can be written in a similar way. From Newton's equations, one derives $\dot{l}' = \dot{x}' \times p' + x' \times \dot{p}' = x' \times \dot{p}' = x' \times F'$ or

$$\dot{l}' = \tau'$$

where τ' represents the torque $\tau' = x' \times F'$. With $l' = Ol$ and $\tau' = O\tau$, one has $l = x \times p$, $\tau = x \times F$ and the equation

$$D_t l = \tau . \tag{4.10}$$

The last relation follows from $D_t l = D_t(x \times p) = (D_t x) \times p + x \times D_t p = x \times D_t p = x \times F = \tau$.

□ *Example: rigid body free motion.*

With the coordinate system attached to a rotating body, $\dot{x} = 0$ for each particle of the body. The angular momentum dl of an element of the body is given by $dl = x \times dp = dm\, x \times D_t x = dm\, x \times (\dot{x} + \Omega x) = dm\, x \times (\Omega x)$. Thus, the equation (4.10) for a free body ($\tau = 0$) can be written in the form

$$\dot{L} + \Omega L = 0 , \quad \text{where} \quad L = \int dm\, (x \times (\Omega x)) . \tag{4.11}$$

The matrix Ω, and thus the vector ϖ, are the same for each particle and therefore can be excluded from integration. Using the vector ϖ, the expression for L can be rewritten as

$$L = \mathcal{I}\varpi ,$$

where $\mathcal{I}$ denotes the tensor of inertia; its Cartesian components are $\mathcal{I}_{ij} = \int dm(x^k x^k \delta_{ij} - x^i x^j)$, where x^k are components of x. Hence, Eq.(4.11) takes the form

$$\mathcal{I}\dot{\varpi} + (\mathcal{I}\varpi) \times \varpi = 0 . \tag{4.12}$$

The matrix $\mathcal{I}$ is symmetric and thus diagonalizable, i.e., there exists a coordinate system in which $\mathcal{I}_{ij} = \mathcal{I}_{(i)}\delta_{ij}$. The explicit form of Eq.(4.12) in that coordinate system is

$$\mathcal{I}_1\,\dot{\varpi}_1+(\mathcal{I}_2-\mathcal{I}_3)\,\varpi_2\,\varpi_3=0, \quad \mathcal{I}_2\,\dot{\varpi}_2+(\mathcal{I}_3-\mathcal{I}_1)\,\varpi_3\,\varpi_1=0, \quad \mathcal{I}_3\,\dot{\varpi}_3+(\mathcal{I}_1-\mathcal{I}_2)\,\varpi_1\,\varpi_2=0.$$

These are the well–known Euler equations. They are easily solved when at least two of $\mathcal{I}_i$ are equal. Let $\mathcal{I}_2=\mathcal{I}_1$. From the third equation, $\dot{\varpi}_3=0$, and ϖ_3 is an integral of motion: $\varpi_3=$ const. The remaining equations can be written as $\dot{\varpi}_1+c\,\varpi_2=0$ and $\dot{\varpi}_2-c\,\varpi_1=0$, where $c=\varpi_3(\mathcal{I}_1-\mathcal{I}_3)/\mathcal{I}_1$. Taking the time derivative of one of the equations and substituting the second one, one obtains the pendulum equations $\ddot{\varpi}_i+c\,\varpi_i=0$, $i=1,2$ with additional initial conditions $\dot{\varpi}_1(0)=-c\,\varpi_2(0)$ and $\dot{\varpi}_2(0)=c\,\varpi_1(0)$. Thus, the most general form of rotation of a mechanically symmetric free rigid body is the regular precession. $\boxtimes$

5

Some Statistical Issues

T HIS chapter contains some topics of practical nature encountered when dealing with large sets of orientations. The first one is related to the calculation of the mean orientation. Later on, a very elementary introduction to some standard distributions on the rotation manifold is given. Moreover, the computer generation of orientations with a foreknown distribution is considered. Finally, there is a very brief section on comparing orientation distributions.

5.1 Mean Orientation

As any measurement of continuous parameters, the determination of orientations is linked with experimental errors. When measuring an orientation, one gets slightly different sets of parameters for each measurement. It is natural to look for a kind of average of such orientations.

Let a set of K orientations be given. The mean orientation must be in a sense 'close' to all orientations of the set. It will be calculated by determining a point at which that 'closeness' given as a function of orientation takes minimum. The distance between orientations is proportional to the angle of the rotation leading from one orientation to another; see Eq.(3.4). In this approach, the minimized quantity is equal to $\sum_{k=1}^{K} w_{(k)}(\omega_{(k)})^2$, where $\omega_{(k)} = \omega(O, O^{(k)})$ $(0 \leq \omega_{(k)} \leq \pi)$ is the misorientation angle between the k-th orientation $(O^{(k)})$ of the set and the sought orientation O, and $w_{(k)}$ is a positive weight corresponding to k-th orientation. There is no closed–form expression for O minimizing $\sum w_{(k)}(\omega(O, O^{(k)}))^2$. The problem can be solved by numerical non–linear optimization; see Barton & Dawson, 2001.

'Geometric' mean
From (3.29), we have the relation $2(\omega(O, O^{(k)}))^2 = \| \log((O^T O^{(k)})) \|^2$ valid for $\omega(O, O^{(k)}) < \pi$. In this case, the minimization of $\sum_{k=1}^{K} (\omega(O, O^{(k)}))^2$ is equivalent to the minimization of

$$\sum_{k=1}^{K} \| \log((O^T O^{(k)})) \|^2 \ .$$

This is similar to the ordinary geometric mean $\bar{x} = (\prod_{k=1}^{K} x_{(k)})^{1/K}$ of positive numbers $x_{(k)}$ which minimizes the sum of squared distances

$$|\ln(x^{-1}x_{(k)})| \ .$$

For more on this subject see Moakher (2002). ⊠

The problem becomes linear, and thus easy to solve, with the approximation $\omega_{(k)}^2 \approx 3 - O_{ij}O_{ij}^{(k)}$. From a different point of view, this is equivalent to the minimization of the weighted sum of squared distances $\chi_{\bullet}$ defined by (3.5). Two equivalent ways of calculating this kind of mean orientation are given below. One is based on orthogonal matrices and the other – on unit quaternions. In terms of orthogonal matrices, the problem is reduced to the polar decomposition of a 3×3 matrix. In terms of quaternions, the solution is an eigenvector of a 4×4 matrix. Both procedures may be confined to proper rotations, or alternatively, in both proper and improper rotations may be allowed. However, the method based on quaternions is more suitable for proper rotations, and it is easier to include improper rotations by using matrices. The methods are presented for finite sets of orientations, but they can be reformulated for continuous distributions of orientations.

The procedures are consistent in the following sense: if all original orientations of the set are rotated by a certain rotation, the resulting mean orientation is rotated by that rotation with respect to the mean of the original orientations. Moreover, if all orientations of the set are equal to a certain orientation, the mean is also equal to that orientation.

5.1.1 Procedure based on orthogonal matrices

Let the orientations be represented by proper orthogonal matrices $O^{(k)}$, $k = 1, 2, ..., K$. A proper orthogonal matrix O^M is sought, for which the sum of squares of distances $\chi_{\bullet}(O, O^{(k)})$ is minimal. The mean orientation O^M is calculated by minimization of the function

$$f(O) = \sum_{k=1}^{K} w_{(k)}(\chi_{\bullet}^2(O, O^{(k)}) - 3) = -\sum_{k=1}^{K} w_{(k)}O_{ij}O_{ij}^{(k)} = -O_{ij}R_{ij} \ , \quad (5.1)$$

where $R_{ij} = \sum_{k=1}^{K} w_{(k)}O_{ij}^{(k)}$ (Humbert $et\ al.$, 1996, Moakher, 2002).

⊡ *The best orthogonal matrix approximating a given matrix.*
A certain matrix R^a is supposed to represent an orientation but it is not exactly orthogonal. The problem is what is the best orthogonal matrix O^M approximating R^a. Of all orthogonal matrices, O^M must be in some sense the 'closest fit' to R^a. A simple choice of the function to be minimized is

$$\|O - R^a\|^2/2 = (O_{ij} - R_{ij}^a)(O_{ij} - R_{ij}^a)/2 = const - O_{ij}R_{ij}^a \ . \quad (5.2)$$

Neglecting the constant term, the function is of the same form as f in (5.1) and analogous means can be applied to determine the point of its minimum. ⊠

⊡ *Procrustes problem: the best rotation relating two sets of vectors*
The problem of the best orthogonal matrix approximating a given matrix appears also in the search for the best rotation relating two sets of vectors. Vectors $x_{(k)}$ and $y_{(k)}$ $(k = 1, ..., K)$ are assumed to be approximately related by

$$y_{(k)} \approx O x_{(k)} \quad \text{or} \quad Y \approx OX , \tag{5.3}$$

where O is an orthogonal matrix, and vectors $x_{(k)}$ and $y_{(k)}$ constitute k–th columns of $3 \times K$ matrices X and Y, respectively. Vectors $y_{(k)}$ should be as close as possible to the rotated vectors $x_{(k)}$, i.e., to $O x_{(k)}$. Again, the idea is to obtain the sought rotation by calculating the point O^M at which a certain function (of O) representing the deviation between these vectors takes minimum. In particular, the matrix O^M can be calculated by determining minima of

$$\sum_{k=1}^{K} w_{(k)}(O x_{(k)} - y_{(k)}) \cdot (O x_{(k)} - y_{(k)}) + const = \\ - \sum_{k=1}^{K} w_{(k)} y_{(k)} \cdot (O x_{(k)}) = - \sum_{k=1}^{K} w_{(k)} O_{ij} Y_{ik} X_{jk} = -O_{ij} R_{ij}^b , \tag{5.4}$$

where $R_{ij}^b = \sum_{k=1}^{K} w_{(k)} Y_{ik} X_{jk}$; MacKenzie (1957), McLachlan (1972), Wahba, Farrell, & Stuelpnagel (1966), Kabsch (1976, 1978) or Stephens (1979) are only a few of numerous papers on this subject.

The point is, that in this case, analogously to the case of the mean rotation, the problem is to find an orthogonal matrix 'closest' (in the sense of the distance (5.2)) to one of R matrices (i.e., R, R^a or R^b).

It is worth to mention Mackay's (1977) idea to approximate O in (5.3) using $O \approx Y X^+$, where X^+ stands for the generalized inverse[1] of X. In general, the matrix $Y X^+ = R^a$ is not exactly orthogonal. Its orthogonal approximation is obtained from the minimization of (5.2). ⊠

The orthogonality conditions for O with $\det(O) = +1$ can be written in the form

$$\det(O) O_{li} O_{lj} = +\delta_{ij} , \tag{5.5}$$

which allows us to distinguish the proper case with $\det(O) = +1$ from the general one in which $\det(O)$ may be equal to $+1$ or -1. With a symmetric matrix Λ of Lagrange multipliers, the function g involving the conditions (5.5) has the form

$$g(O) = f(O) + \Lambda_{ij}(\det(O) O_{li} O_{lj} - \delta_{ij}) .$$

At a stationary point O^M, there occurs $(\partial g/\partial O_{ij})(O^M) = 0$. Hence, from $O_{li}(\partial \det(O)/\partial O_{lj}) = \det(O)\delta_{ij}$,

$$O_{li}^M R_{lj} = \det(O^M) O_{lm}^M O_{lk}^M (\delta_{ij}\Lambda_{mk} + 2\delta_{ik}\Lambda_{jm}) .$$

The orthogonality conditions (5.5) lead to

$$O_{li}^M R_{lj} = L_{ij} , \tag{5.6}$$

where $L_{ij} = (\delta_{ij}\Lambda_{ll} + 2\Lambda_{ij})$. Because of the orthogonality of O^M there occurs

---

[1] The generalized inverse of a matrix X is the unique matrix X^+ satisfying the conditions $X X^+ X = X$, $X^+ X X^+ = X^+$, $(X X^+)^T = X X^+$ and $(X^+ X)^T = X^+ X$ (Penrose, 1955).

$$L^2 = R^T R . \tag{5.7}$$

This equation determines L with accuracy to the signs of its eigenvalues. Moreover, due to (5.6) L must satisfy $\det(L) = \det(R)$. This condition constitutes additional limit for L. The value of f at the point O^M is directly related to the trace of L

$$f(O^M) = -O_{ij}^M R_{ij} = -L_{ii} .$$

The function f takes global minimum for L with maximal trace and, additionally, with the same sign of the determinant as the determinant of R. If $\det(R) > 0$ the residue takes absolute minimum for positive definite L. For $\det(R) < 0$ a proper rotation is obtained if the eigenvalue (of L) with the smallest absolute value is negative, whereas the other two are positive. If $\det(R) = 0$, the undetermined components of the matrix O^M can be set in such a way that the condition $\det(O^M) = +1$ is satisfied; if only one eigenvalue of L vanishes that matrix is unique, otherwise it is not.

⊡ *Improper rotations allowed*

In a more general approach, the rotations (and the matrices) are not necessarily proper, and the interpretation related to $\omega_{(k)}$ must be omitted. In this case, the residue is defined by the last part of (5.1) $f(O) = -\sum_{k=1}^{K} w_{(k)} O_{ij} O_{ij}^{(k)} = -O_{ij} R_{ij}$. With the result O^M allowed to be improper rotation, the standard orthogonality conditions $O^T O = I_3$ are used. The minimized function g involving the conditions has the form $g(O) = f(O) + \Lambda_{ij}(O_{li} O_{lj} - \delta_{ij})$ and at the stationary points

$$R_{ij} = 2O_{il}^M \Lambda_{lj} .$$

The orthogonality conditions lead to

$$O_{li}^M R_{lj} = L_{ij} ,$$

where $L_{ij} = 2\Lambda_{ij}$. As in the case limited to proper rotations there occurs

$$L^2 = R^T R , \quad \text{and} \quad f(O^M) = -L_{ii} .$$

With the mean rotation allowed to be proper or improper, the residue f takes absolute minimum for L with maximal trace, i.e., for L with largest eigenvalues. This means that from possible solutions of $L^2 = R^T R$, the positive semi-definite matrix L must be chosen. If R is not singular the condition (5.6) determines O^M uniquely. If $\det(R) = 0$ the solution is not unique but positive semi-definiteness of L still guarantees minimal value of f for O^M satisfying (5.6). In other words, this case is solved by the polar decomposition of R (Halmos, 1958). ⊠

⊡ *'Harmonic' mean equals 'arithmetic' mean*

If O^M obtained via (5.6) is the mean of the set of orientations $O^{(k)}$ $((k) = 1, ..., K)$ then $(O^M)^T$ is the mean of the set of orientations $(O^{(k)})^T$. Due to the orthogonality of the matrices, analogous statement is true for $(O^M)^{-1}$ and $(O^{(k)})^{-1}$. Thus, on $SO(3)$, the analogue of the 'harmonic' averaging method leads to the same result as the 'arithmetic' approach. ⊠

5.1.2 Procedure based on quaternions

The method given in the previous section can be reformulated in terms of quaternions. It is assumed that the orientations to be averaged are represented by quaternions $q_{(k)}$, $k = 1, 2, ...K$. Quaternion q_M corresponding to an orientation minimizing the weighted sum of squared distances $\chi_\bullet$ is sought. The relation between an orthogonal matrix O of proper rotation and corresponding quaternion q is given by (2.43) $O_{ij} = \mathsf{G}_{ij\mu\nu}q^\mu q^\nu$. The quantity $-\sum_{k=1}^{K} w_{(k)}O_{ij}O_{ij}^{(k)}$ which defines the residue f in Eq.(5.1) is equal to $-\sum_{k=1}^{K} w_{(k)}\mathsf{G}_{ij\iota\kappa}\mathsf{G}_{ij\mu\nu}q^\iota q^\kappa q_{(k)}^\mu q_{(k)}^\nu$. Taking into account the relation $\mathsf{G}_{ij\iota\kappa}\mathsf{G}_{ij\mu\nu} = 2(\delta_{\iota\mu}\delta_{\kappa\nu} + \delta_{\iota\nu}\delta_{\kappa\mu}) - \delta_{\iota\kappa}\delta_{\mu\nu}$ and the fact that $q_{(k)}$ are of unit magnitude, the residue can be expressed as

$$f_1(q) = -P_{\mu\nu}q^\mu q^\nu , \tag{5.8}$$

where P is a traceless symmetric matrix defined by $P_{\mu\nu} = \sum_k w_{(k)}(4q_{(k)}^\mu q_{(k)}^\nu - \delta_{\mu\nu})$. The quaternion q must have unit magnitude, i.e., $q^\mu q^\mu - 1 = 0$. Let

$$g_1(q) = f_1(q) + \lambda(q^\mu q^\mu - 1) .$$

The stationarity condition $(\partial g_1/\partial q^\mu)(q_M) = 0$ gives

$$Pq_M = \lambda q_M . \tag{5.9}$$

Moreover, there occurs

$$f_1(q_M) = -\lambda . \tag{5.10}$$

Thus, the residue is minimized by the normalized eigenvector q_M corresponding to the largest eigenvalue of P. The rotation is unique provided this eigenvalue is not degenerate.

The formalism presented above is analogous to the quaternion method of calculating the best rotation relating two sets of vectors which was independently presented by Faugeras & Hebert (1983), Horn (1987) and Diamond (1988, 1990). See also Morawiec (1998).

> **Improper rotations allowed**
> Only proper rotations are taken into account in the above procedure. However, it can be extended to improper rotations. Let us assume that the k-th rotation is improper. The corresponding orthogonal matrix $O^{(k)}$ can be expressed as $\mathfrak{I}\underline{O}^{(k)}$, where $\mathfrak{I}$ represents inversion ($\mathfrak{I}_{ij} = -\delta_{ij}$) and $\underline{O}^{(k)}$ is a proper orthogonal matrix. With correspondence between $\underline{O}^{(k)}$ and quaternion given by $\underline{O}_{ij}^{(k)} = \mathsf{G}_{ij\mu\nu}q_{(k)}^\mu q_{(k)}^\nu$, one has $O_{ij}^{(k)} = -\mathsf{G}_{ij\mu\nu}q_{(k)}^\mu q_{(k)}^\nu$. Thus, the improper rotation can be included by finding the quaternion corresponding to that rotation composed with inversion, and by replacing $w_{(k)}$ by $-w_{(k)}$ in the sum constituting P.

In order to allow the mean rotation to be improper, O is replaced by $\mathfrak{I}\underline{O}$ with $\det(\underline{O}) = +1$. The right–hand side of (5.8) changes its sign; thus, we look for the location of the maximum of f_1. With the relations (5.9) and (5.10) still valid, the function is maximized if λ is the smallest eigenvalue of P. The sought improper rotation is equal to the inversion composed with the proper rotation represented by the eigenvector corresponding to that eigenvalue. Whether the final result is proper or improper rotation depends on what is larger: the largest eigenvalue of P or the absolute value of its smallest eigenvalue. $\boxtimes$

□ *Another approach based on quaternions.*

If quaternions q and $q_{(k)}$ represent two orientations separated by the distance $\omega_{(k)}$, then $\cos(\omega_{(k)}/2) = q^\mu q_{(k)}^\mu$. For small $\omega_{(k)}$, $\cos(\omega_{(k)}/2) \approx 1 - \omega_{(k)}^2/8$; thus, instead of $\sum_{k=1}^K w_{(k)} \omega_{(k)}^2$, one can minimize

$$-\sum_{k=1}^K w_{(k)} \cos(\omega_{(k)}/2) = -\sum_{k=1}^K w_{(k)} q^\mu q_{(k)}^\mu = -q^\mu r_\mu , \qquad (5.11)$$

where $r_\mu := \sum_{k=1}^K w_{(k)} q_{(k)}^\mu$. Taking into account the condition $q^\mu q^\mu - 1 = 0$, after differentiating the expression $-q^\mu r_\mu + \lambda(q^\mu q^\mu - 1)/2$ with respect to q^μ and setting the result to be zero, we obtain $r_\mu = \lambda q^\mu$. At the point q satisfying this relation, the residue is given by $-q^\mu r_\mu = -\lambda q^\mu q^\mu = -\lambda$. On the other hand, the multiplier λ satisfies $\lambda^2 = r_\mu r_\mu$ and hence, the expression (5.11) takes minimum for $\lambda = +(r_\mu r_\mu)^{1/2}$. Thus, for non-zero r, $q^\mu = r_\mu/(r_\nu r_\nu)^{1/2}$. Of all unit quaternions, $r_\mu/(r_\nu r_\nu)^{1/2}$ is the closest to r in the sense of the Euclidean distance in 4 dimensional vector space; this follows from basic geometrical considerations.

This procedure can be applied to find the average of proper rotations but there is an additional difficulty which must be taken into account. The correspondence between quaternions and proper rotations is two–to–one, and quaternions with opposite signs correspond to the same rotation. Therefore, not just one $r_\mu = \sum_{k=1}^K w_{(k)} q_{(k)}^\mu$ but all cases $\sum_{k=1}^K w_{(k)}(\pm q_{(k)}^\mu)$ with various combinations of signs must be considered. The one providing the minimal value of the residue, i.e., the one with maximal $\lambda = (r_\mu r_\mu)^{1/2}$, is the correct solution. This means that the final r and λ are always non–zero and the procedure gives a unique solution unless the largest value of λ corresponds to two or more different combinations. ("Different" means that the sums r are neither equal nor mutually opposite vectors.)

In general, the method based on (5.11) leads to results different than those obtained from minimization of (5.1) and (5.8). The distance between them can be significant if the orientations are scattered. $\boxtimes$

5.2 Distributions on the Rotation Manifold

One of the main subjects of texture analysis is the distribution of orientations.[2] Experimentally obtained distributions, say displayed in the Euler angle parameterization, usually have the form of "valleys and hills", sometimes "ranges of hills". Elevated areas are called texture components; ranges are referred to as texture fibers. A component is characterized by its location and shape, and it is desirable to be able to specify that characterization by a small number of parameters. The standard approach is to fit a function of a given form to the texture component. The choice of the model function, however, is not trivial. It must suit the topology of $SO(3)$ and be flexible enough to fit various shapes of components. On the other hand, we would like it to be

---

[2] The function referred to as "distribution" in texture analysis would be called "a probability density function" by a mathematician.

relatively simple. The issue of distributions on the rotation manifold is closely related to distributions of directional data. For more on the latter subject see Mardia (1972).

In the case of data given on the axis of real numbers (or, more generally, in R^n), there is one distribution with an outstanding position – the (Gauss) normal distribution. Its prominence follows from the role of the normal distribution in the so called central limit theorem, and also from a number of other characterizations which make it special. It must be stressed that there is no such special case among orientation distributions. This is illustrated below on a simple 2D case.

□ *Normal distribution*
The one dimensional normal distribution $N(\mu, \sigma)$ has the density function $(2\pi\sigma)^{-1/2} \exp\left((x - \mu)^2/(2\sigma)\right)$. The mean value and variance are $E(x) = \mu$ and $E\left((x - \mu)^2\right) = \sigma$, respectively. Let us list a few of its numerous appearances.

Central limit theorem: Let x_i be a series of independent random variables with the same distribution and the mean μ and variance σ. The variable $S_m = \sum_{i=1}^{m} x_i$ has the mean value $m\mu$ and variance $m\sigma^2$. The limit distribution of the series of the variables $\zeta_m = (S_m - m\mu)/(\sqrt{m}\sigma)$ is that corresponding to $N(0, 1)$.

The maximum likelihood: For a given probability density function f, the likelihood function is defined as $\prod_{i=1}^{m} f(x_i - \bar{x})$. The sample mean $\bar{x} := \frac{1}{m}\sum_{i=1}^{m} x_i$ is the maximum likelihood estimate of the true mean value if and only if the distribution of the random variable x is normal.

The maximum entropy: With fixed mean value and variance, the entropy $-\int_{-\infty}^{+\infty} f(x) \log f(x) dx$ is maximized when f is the density function of the normal distribution.

Brownian motion: A particle is at position $x = 0$ at time $t = 0$. Then it moves infinitesimal distances in infinitesimal periods of time. The probability of the particle's position at the time t is described by the normal distribution.

Diffusion equation: The normal distribution is the solution of the diffusion equation

$$D(\partial^2 f/\partial x^2) = \partial f/\partial t \ , \quad \text{where} \quad f = f(x, t) \ ,$$

with constant D and the initial condition $f(x, 0) = \delta(x)$, i.e., with all "mass" concentrated at $x = 0$ at the outset of the diffusion process. ⊠

□ *Distributions of 2D rotations*
One can consider analogous statements for data which are not from R^n. In the case of orientations, those special properties are divided between a number of distributions. To see that and as a kind of introduction, orientations on a plane will be considered.

An orientation in two dimensions is specified by just one angle θ from the range $[0, 2\pi)$ and can be identified with a point of the unit circle. Let the distribution be described by $f(\theta)$. Obviously, the uniform distribution is given by $f(\theta) = (2\pi)^{-1}$, $(\theta \in [0, 2\pi))$. If the one dimensional normal distribution $N(0, \sigma^2)$ is wrapped around the circle, the so–called *wrapped normal* distribution is obtained; the explicit expression for the density function is

$$f(\theta) = \frac{1}{2\pi}\left(1 + 2\sum_{p=1}^{\infty}\rho^{p^2}\cos(p\theta)\right) \ , \ \text{where} \ \ \rho = e^{-\sigma^2/2} \ \ \ \theta \in [0, 2\pi] \ .$$

It is intuitively obvious that the wrapped normal distribution describes "Brownian motion" of orientations in two dimensions. However, in this case there is another distinct function – the von Mises distribution

$$f(\theta) = \frac{1}{2\pi I_0(\kappa)}e^{\kappa\cos\theta} \ , \ \ \ \kappa > 0 \ , \ \ \ \theta \in [0, 2\pi] \ .$$

The normalization coefficient involves the zero order modified Bessel function of the first kind $I_0(\kappa) = \sum_{r=0}^{\infty}(r!)^{-2}(\kappa/2)^{2r}$. The distinctiveness of the von Mises distribution follows from the fact that the entropy of the rotational data is maximized for this distribution. Moreover, the mean orientation μ is the maximum likelihood estimate of true mean iff the distribution of the random variable θ is the von Mises distribution. The central limit theorem involves the uniform distribution. $\boxtimes$

5.2.1 Von Mises–Fisher distribution

The von Mises distribution (and its counterpart on the sphere – Fisher distribution) can be given for an arbitrary dimension. In general, the orientation distribution of p-dimensional objects in N-dimensional space ($p \le N$) can be considered. As was explained in chapter 1, objects' orientations can be specified by sets of p mutually orthogonal unit vectors $r^{(i)}$, $i = 1, ..., p$. Let X be the $p \times N$ matrix built of $r^{(1)}, ..., r^{(p)}$, i.e.,

$$X = \begin{bmatrix} r^{(1)T} \\ \vdots \\ r^{(p)T} \end{bmatrix} . \tag{5.12}$$

Due to the orthonormality of vectors, there occurs $XX^T = I_p$. The probability density function of the generalized von Mises–Fisher distribution is

$$f_{vMF}(X; F) = (1/c(F))\exp\left(\text{tr}(FX^T)\right) , \tag{5.13}$$

where F is a $p \times N$ parameter matrix (Downs, 1972). The normalization coefficient $c(F)$ can be expressed by hypergeometric function $_0F_1$ of matrix argument: $c(F) = {_0F_1}(N/2, FF^T/4)$; in fact, instead of FF^T, the coefficient $c(F)$ is fully determined by the eigenvalues of the matrix FF^T (Khatri & Mardia, 1977).

Let us limit our considerations to the case most important for crystallographic textures with $N = 3 = p$ and $\det(X) = +1$. F as a square matrix can be decomposed into the product KO^a, where K is symmetric and O^a is proper orthogonal. If $F = 0$, the corresponding distribution is uniform. If K is symmetric and positive definite, the distribution has a maximum at the point O^a; for negative definite K, the distribution has a minimum at that point. With $K = \alpha I_3$, the distribution takes the form

$$(1/c(\alpha^2 I_3))\exp\left(\text{tr}(\alpha O^a X^T)\right) \propto \exp(2\alpha\cos(\omega)) ,$$

where ω represents misorientation angle between O^a and X, i.e., the value of the density function depends only on the distance from O^a.[3]

5.2.2 Bingham distribution

As was noticed in the chapter on parameterizations, the unit quaternions representing orientations are located on the three–dimensional sphere S_3 in four–dimensional space. Now, the idea is to use a distribution given on the sphere as the orientation distribution. Such distribution, however, must be centro–symmetric because q and $-q$ correspond to the same rotation. We already have a distribution on the sphere in the form of the von Mises–Fisher distribution with $p = 1$ and $N = 4$ but, in general, it is not centro–symmetric. The condition of centro–symmetry is satisfied when the exponent contains an even function of the unit vector q on the 3–sphere; the simplest one is the quadratic expression $q^T A q$, where A is a matrix. Since the antisymmetric part of A does not influence $q^T A q$, the matrix A is assumed to be symmetric. The distribution proportional to $\exp(q^T A q)$ is called Bingham distribution. With A decomposed into $A = \mu \kappa \mu^T$, where μ is an orthogonal 4×4 matrix and κ is a diagonal matrix, the Bingham distribution is usually written in the form

$$f(q; \kappa, \mu) = (1/c(\kappa)) \exp\left(q^T \mu \kappa \mu^T q\right) , \quad q^T q = 1 , \quad \mu^T \mu = I_4 , \quad (5.14)$$

As was assumed, $f(-q; \kappa, \mu) = f(q; \kappa, \mu)$. It is also easy to see that the transformation $\kappa \to \kappa + \lambda I_4$ does not change the distribution; to get rid of that ambiguity, a condition involving κ is added (e.g. $\text{tr}(A) = \text{tr}(\kappa) = 0$). The normalizing coefficient turns out to depend only on κ and is given by a confluent hypergeometric function of matrix argument $c(\kappa) = {}_1F_1(1/2, 2, \kappa)$ (see, e.g., Prentice, 1986 or Schaeben, 1996). As for the characterization of the distribution, entropy and likelihood on the hemisphere S_+^3 reach maxima when the density function is given by (5.14).

However, the crucial point is that the von Mises-Fisher and Bingham orientation distributions are actually identical: if proper orthogonal matrices have the von Mises-Fisher distribution, the corresponding quaternions have the Bingham distribution (Prentice, 1986). With the relation of proper orthogonal matrices and unit quaternions written as $X_{ij} = \mathsf{G}_{ij\mu\nu} q^\mu q^\nu$ (Eq.2.43), the density function of the von Mises-Fisher distribution involves the exponent of $\text{tr}(FX^T) = F_{ij} X_{ij} = F_{ij} \mathsf{G}_{ij\mu\nu} q^\mu q^\nu$, i.e., it has the form of Bingham distribution. The matrix A is related to F via $A_{\mu\nu} = F_{ij} \mathsf{G}_{ij\mu\nu}$.

5.2.3 Other distributions

There are a number of other orientation distributions applicable in texture analysis. Many of them depend only on a distance ω from a fixed orientation, say O^a; such distributions are referred to as 'central'. For simplicity, they are expressed as functions of ω; by using $\omega = \arccos((\text{tr}(O^a X^T) - 1)/2)$, the functions depending directly on orientations X can be easily obtained.

---

[3] Such function was called "Gauss–shaped standard function" by Matthies et al. (1987).

Savyolova (1984) gave a distribution which appears in an analogue of the central limit theorem valid on $SO(3)$. In the particular central case, it has the form

$$f_{Brown}(\omega; t) = \frac{1}{8\pi^2} \sum_{k=0}^{\infty} F_k(t) \sin((2k+1)\omega/2)/\sin(\omega/2) ,$$

where $F_k(t) := (2k+1)\exp(-k(k+1)t^2)$, and t is a parameter. It is also the distribution of the counterpart of the orientation Brownian motion (Savyolova, 1993). Moreover, the above distribution is the solution of the analogue of the diffusion equation on $SO(3)$ (Bucharova and Savyolova, 1993).

It is worth remembering the function

$$f_L(\omega; t) = (1 - t^2) \frac{(1 + t^2)^2 + 4t^2 \cos^2(\omega/2)}{((1 + t^2)^2 - 4t^2 \cos^2(\omega/2))^2} , \quad \omega \in [0, \pi] , \quad |t| < 1 ,$$

where t is a parameter and ω is the misorientation angle between a given orientation and a fixed mean orientation. More on this function can be found in the book by Matthies et al. (1987), where it is referred to as the "Lorentz–shaped standard function".

Let us also mention the de la Vallée Poussin distribution. Its value as a function of the distance ω from a fixed orientation is

$$f_{VP}(\omega; k) = \sqrt{\pi} \frac{\Gamma(k+2)}{\Gamma(k+1/2)} \cos^{2k}(\omega/2) ,$$

where the parameters k are non–negative integers, and Γ is the Euler gamma function. The advantage of this simple function is that it has a finite series expansion. For more details, see Schaeben (1997).

5.3 Generation of Orientations

The practical issue of sampling of the rotation space is fundamental for texture related computer simulations. A simulated model material is expected to be similar to its real archetype. This includes the distribution of orientations. Thus, one faces the necessity to create sets of orientations with a postulated distribution.

5.3.1 Random orientations

The first basic question is how to generate a set of orientations with uniform distribution. That particular case is usually referred to as the generation of *random* orientations. The procedure must involve the invariant volume element for the parameterization in which the generation is performed. Three simple algorithms are described below. All require a random number generator, i.e., a piece of software which will give random numbers in the range, say $[0, 1]$.

The first algorithm based on the Euler angles is most commonly used. Random orientations are generated by taking three random numbers within $[0, 1]$ as $\varphi_1/(2\pi)$, $(\cos \phi + 1)/2$ and $\varphi_2/(2\pi)$.

Algorithm 1

The step by step procedure is:

1. Generate three random numbers x_i, $(i = 1, 2, 3)$ in the range $[0, 1]$ each.
2. Transform the first two by $\varphi_i = 2\pi x_i$, $(i = 1, 2)$ to the range $[0, 2\pi]$.
3. Transform the third one by $\phi = \arccos(2x_3 - 1)$ to the range $[0, \pi]$.
4. Consider the numbers $\varphi_1, \phi, \varphi_2$ to be the Euler angles of the randomly generated orientation.
5. Go to point 1 to generate the next orientation. ☒

One could also follow the appealing concept of using the isochoric parameterization ρ^i with $\rho^i = n_i(3(\omega - \sin\omega)/(4\pi^2))^{1/3}$; see section 2.7, Eq.(2.54). Because of the trivial form of invarinat volume element, random orientations would be generated by taking random numbers within $[0, 1]$ as $(2\rho^i + 3)/6$. If $\sqrt{\rho^i\rho^i}$ is smaller than $(3/(4\pi))^{1/3}$ the triplet would be accepted as isochoric parameters; otherwise, it would be discarded.

Algorithm 2

The step by step procedure is:

1. Generate three random numbers x_i $(i = 1, 2, 3)$ in the range $[0, 1]$ each.
2. Transform them by $\rho^i = 2ax_i - a$ to the range $[-a, a]$, where $a = (3/(4\pi))^{1/3}$.
3. If $\rho^i\rho^i$ is larger than a^2, go to step 1.
4. Otherwise, consider the resulting triplet ρ^i to be isochoric parameters.
5. To generate the next orientation, go to point 1. ☒

The disadvantage of the above algorithm is that there is no simple way to get other parameters from the isochoric parameters, and the method is useless from the practical viewpoint.

However, as was already mentioned, the space of isochoric parameters is related to the quaternion hemisphere via equal–volume projection. This means that uniformly distributed orientations are represented by quaternions uniformly distributed on the hemisphere. Therefore, random orientations can be obtained by generating points randomly distributed on the unit quaternion sphere (van den Boogaart, 2002).

Algorithm 3

The procedure is:

1. Generate four random numbers x_μ, $(\mu = 0, ..., 3)$ in the range $[-1, 1]$ each.
2. Calculate $|x|^2 = x_\mu x_\mu$.
3. If $|x|^2$ is out of the range $[\epsilon^2, 1]$, where $\epsilon^2 < 1$ is a fixed small number, go to 1.
4. Consider the numbers $q^\mu = x_\mu/|x|$ to be components of a quaternion corresponding to the generated orientation.
5. Go to point 1 to generate the next orientation. ☒

5.3.2 Non–uniform distributions

The generation of randomly distributed orientations is useful when testing theoretical ideas. However, in simulations based on real textures, we would like to be able to generate orientations with a given *non–uniform* distribution.

In order to write a procedure generating orientations with an arbitrary distribution f, a routine generating orientations with uniform distribution is needed. The task can be performed by general 'rejection sampling' which works by generating random variables and then rejecting some of them with a probability depending on the value of orientation density. A randomly generated orientation O is accepted if the value of f at O is not larger than a random number from the range $[0, \max(f)]$.

Rejection algorithm
More precisely, the following steps must be taken:
1. Determine the maximum $\max(f)$ of the considered distribution f.
2. Generate a random number x (in the range $[0, 1]$).
3. Using the procedure given in the previous subsection, generate an orientation O corresponding to the uniform distribution.
4. If $x \max(f) > f(O)$ discard O; otherwise, accept (save) it.
5. In order to create the next orientation, go to point 2. ⊠

5.4 Comparing Smooth Orientation Functions

Another frequently encountered issue is to determine the level of similarity between two orientation distributions. Without using sophisticated mathematical concepts, simple indicators of similarity can be applied. E.g., the deviation between the distributions f_1 and f_2 can be quantitatively estimated by

$$\frac{1}{2} \int |f_1(O) - f_2(O)| \, \mathrm{d}O \ .$$

Due to normalization of f_1 and f_2, this number is in the range from 0 to 1 with the former value corresponding to (almost everywhere) equal distributions.

The above quantity can be applied in the practical issue of estimating the "texture sharpness"; in this case, a given distribution f is compared with the uniform (random) distribution: $\int |f(O) - 1| \, \mathrm{d}O/2$. In texture analysis, the texture sharpness is usually quantified using the so–called texture index defined by

$$\int f(O)^2 \mathrm{d}O \ ;$$

see, e.g., Bunge, 1982. The texture index is in the range from 1 (uniform distribution) to infinity. Another simple measure of the sharpness of non–vanishing distributions is the statistical entropy

$$-\int f(O) \ln(f(O)) \mathrm{d}O \ . \tag{5.15}$$

The entropy is zero for the uniform distribution and minus infinity for distributions concentrated around one orientation.

6

Symmetry

T HE issue of object's orientation has some additional aspects when the object exhibits a rotational symmetry. Symmetry is a feature of most crystals (and crystallites in a polycrystalline material) and hence, these new aspects are of importance in the analysis of crystallographic textures.

The material of sections 6.1 and 6.2 can be found in an extended form in numerous texts introducing to crystallography. It is outlined here for completeness.

6.1 Finite Point Groups

The symmetries relevant to orientations are associated with displacements. A symmetry operation of a finite object is a displacement such that the object's initial configuration with respect to its environment is identical to the configuration after the displacement. The object is symmetric if it has a non–trivial symmetry operation. It is easy to see that the symmetry operations constitute a group. If there is a point fixed under all operations of the group, the latter is called a point group. The symmetry operations of a point group are rotations.

In three dimensions, point groups can be relatively easily classified. The first step is to enumerate all point groups containing only proper rotations.

- The simplest case concerns an object having no symmetry with a fixed point, i.e., the null displacement is the only element of the point group. This point group is denoted by C_1.
- Rotations about one axis by the angles $2k\pi/n$, $(k = 0, 1, ..., n - 1)$ constitute a group of the order n. It is called a cyclic group and is denoted by C_n. The axis is referred to as an n–fold rotation axis.
- More complex is the case with two or more distinct axes. Symmetry operations transform a given axis into other axes; all of them constitute a class of axes of the same foldness. For the group of order M, the class k of m_k–fold axes contains $M/(2m_k)$ distinct axes.

Explanation
This can be shown by considering a general point close to an axis of class k.

Rotations about that axis transform the point into $2m_k$ points. The factor 2 appears because, with the presence of the intersecting axis, the considered axis is always transformed onto itself with the opposite sense. On the other hand, the complete set of all possible images of the point contains M elements. Hence, the number of axes in class k is $M/(2m_k)$. ☒

The number of symmetry operations other than the identity in the k–th class is $(m_k - 1)M/(2m_k)$. The sum of them is equal to $M - 1$, i.e., to the number of all symmetry operations other than the identity. Thus, $\sum_{k=1}^{K}(m_k - 1)M/(2m_k) = M - 1$ or equivalently

$$2(1 - 1/M) = \sum_{k=1}^{K}(1 - 1/m_k) ,$$

where $K \geq 2$ is the number of the classes. All possible solutions to that equation can be written in the form:
1. $K = 2$, $m_1 = M = m_2$, (C_M),
2. $K = 3$, $m_1 = 2$, $m_2 = 2$, $M = 2m_3$, (D_{m_3}),
3. $K = 3$, $m_1 = 2$, $m_2 = 3$, $m_3 = 3$, $M = 12$, (T),
4. $K = 3$, $m_1 = 2$, $m_2 = 3$, $m_3 = 4$, $M = 24$, (O),
5. $K = 3$, $m_1 = 2$, $m_2 = 3$, $m_3 = 5$, $M = 60$, (I),
and other combinations of the assignment of these values to m_k within each line. As indicated, the first line corresponds to the cyclic group. The second one represents another infinite family of point groups called dihedral groups D_n ($n = 2, 3, ...$). The dihedral group D_n has a symmetry operation with an n–fold principal axis and n twofold axes perpendicular to the principal axis; it is the point symmetry group of prisms with n–sided regular polygons as bases. The remaining three groups are the tetrahedral (T), octahedral (O) and icosahedral (I) groups. As the names indicate, T, O and I are symmetry groups of the regular solids: tetrahedron, octahedron (and cube) and icosahedron (and dodecahedron), respectively. The tetrahedral group T involves four threefold axes perpendicular to the faces of the tetrahedron and three twofold axes perpendicular its edges. The octahedral group O involves three fourfold axes perpendicular to the faces of the cube, four threefold axes through its vertices and six twofold axes perpendicular its edges. The icosahedral group I involves six fivefold axes through the vertices of the icosahedron, ten threefold axes perpendicular to its faces and 15 twofold axes perpendicular its edges.

Another approach
The case with two or more distinct axes can be considered in relation to the Rodrigues–Hamilton theorem (chapter 1). Let the group contain rotations about two distinct m_1–fold and m_2–fold axes. The composition of the rotations about these axes by the angles $2\pi/m_1$ and $2\pi/m_2$ gives a rotation which must belong to the point group. Let that rotation be m_3–fold. By Rodrigues–Hamilton theorem and using the spherical cosine formula, the numbers m_i must satisfy

$$\cos(\pi/m_1)\cos(\pi/m_2) + \cos(k\pi/m_3) = \sin(\pi/m_1)\sin(\pi/m_2)\cos(\nu) ,$$

where k is one of $1, ..., m_3 - 1$, and ν is the angle between the m_1–fold and m_2–fold axes. The possible solutions of this relation

1. $m_1 = 2, 3, 4, ..., m_2 = 2 = m_3$
2. $m_1 = 2, m_2 = 3, m_3 = 3, 4, 5$

(and all other combinations of the assignment of these numbers to m_1, m_2 and m_3) lead to the already listed groups. The first class of finite groups corresponds to the dihedral groups. The second class corresponds to tetrahedral ($m_3 = 3$), octahedral ($m_3 = 4$) and icosahedral ($m_3 = 5$) groups. ☒

Summarizing, the only finite groups of proper rotations in the three dimensional space are the two infinite families of cyclic groups C_n ($n = 1, 2, ...$) and dihedral groups D_n ($n = 2, 3, ...$), and the tetrahedral (T), octahedral (O) and icosahedral (I) groups.

All other finite point groups in three dimensions are obtained by adding improper rotations to C_n, D_n, T, O and I.

- The group C_{nh} is obtained from C_n by adding the reflection with respect to the plane perpendicular to the axis.
- The group C_{nv} is obtained from C_n by adding the reflection with respect to the plane containing the axis. Obviously, C_{1h} is geometrically identical to C_{1v}; it is usually denoted by C_s.
- By adding (to C_n) the element which is the composition of
  i. the rotation by π/n about the axis of C_n and
  ii. the reflection with respect to the plane perpendicular to that axis
  one gets the group denoted by S_{2n}. The group S_2 is also denoted by C_i.
- The group D_{nh} is obtained from D_n by adding the reflection with respect to the plane perpendicular to the principal axis.
- The group D_{nd} is obtained from D_n by adding the reflection with respect to the plane containing the principal axis and bisecting the angle between neighboring twofold axes.
- The group T_d is obtained by adding to T the reflection with respect to the plane containing the twofold and threefold axes.
- The groups T_h, O_h and I_h are obtained by adding inversion to T, O and I, respectively.

From the point of view of the group theory, some of the groups in the above classification are isomorphic. For instance, C_2, C_i and C_s, as groups of order 2, are isomorphic. The classification makes distinction between displacements and is referred to as the geometric classification. A more precise base for the geometric classification is obtained by representing symmetry operations, by orthogonal matrices. Let the 3×3 orthogonal matrices S_i, S_i' represent symmetry operations of point groups G and G', respectively. G and G' are said to belong to the same geometric class if there is an orthogonal matrix L such that $LS_iL^{-1} = S_i'$ for all elements of the groups. This means that the geometric classification is independent of the choice of the basis.

6.2 Crystallographic Point Groups

A point group of a lattice is called a crystallographic point group. Its operations bring the lattice into self-coincidence. A crystallographic point group has a finite order.

Evidence

Take a finite ball centered at the fixed point and containing basis vectors of the lattice basis. The ball contains a finite set of lattice points and symmetry operations permute these points. The number of permutations of a finite set is finite; hence, the number of symmetry operations is finite.

It is worth mentioning that the following related statement can be proved: if symmetry operations of an object constitute a group of *finite order*, there is a point fixed under all these operations. ⊠

For a given lattice, the point symmetries can be determined by using the condition (2.57) for the rotation matrix R in a non-Cartesian coordinate system. Rotations corresponding to symmetry operations transform lattice points on lattice points. Thus, the entries of R expressed in the lattice basis must be integers. Hence, the symmetry operations correspond to integer matrices satisfying $g_{ij} R^i{}_k R^j{}_l = g_{kl}$, where g is a metric of the lattice.

Cubic case O_h

Let the metric be given by $g_{ij} = \delta_{ij}$. The condition (2.57) takes the form $R^i{}_k R^i{}_l = \delta_{kl}$. Thus, R is an orthogonal matrix with integer entries. The only possible entries are 0 and ± 1 and there can be only one non–zero entry (i.e., $+1$ or -1) per column and per row; it is easy to construct all 48 different matrices satisfying these rules. ⊠

Hexagonal case D_{6h}

Let the metric be given by

$$g = a^2 \begin{bmatrix} 1 & -1/2 & 0 \\ -1/2 & 1 & 0 \\ 0 & 0 & (c/a)^2 \end{bmatrix}. \tag{6.1}$$

The conditions $g_{ij} R^i{}_k R^j{}_l = g_{kl}$ are satisfied by

$$\begin{bmatrix} R^1{}_1 & R^1{}_2 \\ R^2{}_1 & R^2{}_2 \end{bmatrix} =$$

$$\begin{bmatrix} 1 & 0 \\ 0 & 1 \end{bmatrix}, \begin{bmatrix} -1 & 0 \\ 0 & -1 \end{bmatrix}, \begin{bmatrix} 1 & -1 \\ 1 & 0 \end{bmatrix}, \begin{bmatrix} -1 & 1 \\ -1 & 0 \end{bmatrix}, \begin{bmatrix} 0 & 1 \\ -1 & 1 \end{bmatrix}, \begin{bmatrix} 0 & -1 \\ 1 & -1 \end{bmatrix},$$

$$\begin{bmatrix} 0 & 1 \\ 1 & 0 \end{bmatrix}, \begin{bmatrix} 0 & -1 \\ -1 & 0 \end{bmatrix}, \begin{bmatrix} -1 & 1 \\ 0 & 1 \end{bmatrix}, \begin{bmatrix} 1 & -1 \\ 0 & -1 \end{bmatrix}, \begin{bmatrix} 1 & 0 \\ 1 & -1 \end{bmatrix}, \begin{bmatrix} -1 & 0 \\ -1 & 1 \end{bmatrix},$$

with $R^1{}_3 = R^2{}_3 = R^3{}_1 = R^3{}_2 = 0$ and $R^3{}_3 = \pm 1$. This gives 24 symmetry operations of the hexagonal lattice in the (non–Cartesian) coordinate system based on vectors a_i such that $a_i \cdot a_j = g_{ij}$. ⊠

The trace of R, which is equal to $1 + 2\cos\omega$ (Eq.(2.59)), is also an integer. Hence, the cosine of the rotation angle must take the values $\cos\omega = 0, \pm 1/2, \pm 1$ for elements of crystallographic groups. Thus, only 1, 2, 3, 4 and 6–fold rotation axes are allowed (the well–known condition referred to as the crystallographic restriction).

Point groups containing only proper rotations are referred to as point groups of the first kind. Those containing improper rotations are point groups of the second kind. There are 11 crystallographic point groups of the first kind:

C_1, C_2, C_3, C_4, C_6, D_2, D_3, D_4, D_6, T and O. Groups of the second kind obtained from the groups of the first kind by adding the inversion are C_i, C_{2h}, S_6, C_{4h}, C_{6h}, D_{2h}, D_{3d}, D_{4h}, D_{6h}, T_h and O_h, where C_i denotes the group consisting of the identity and the inversion; in the context of diffraction, they are usually referred to as 'Laue groups'. The remaining groups of the second kind involve reflections but no inversion; there are 10 of them C_s, S_4, C_{3h}, C_{2v}, C_{3v}, C_{4v}, C_{6v}, D_{2d}, D_{3h} and T_d. In total, there are 32 geometric classes of crystallographic point groups.

The point group containing all fixed point symmetry operations of a lattice is called the holohedry of that lattice. Two lattices belong to the same crystallographic system if their holohedra belong to the same geometric class. In three dimensional space there are seven crystallographic systems. The classification of point groups into crystallographic systems is given in Table 6.1.

Table 6.1. The crystallographic systems and the classification of point groups. The first column indicates the presence of inversion and reflections: p – groups containing proper rotations only, i – groups containing central inversion, m – groups of second kind containing reflections but no inversion. Holohedra are listed in the first row.

	Triclinic	Monoclinic	Orthorhombic	Trigonal	Tetragonal	Hexagonal	Cubic
i	C_i	C_{2h}	D_{2h}	D_{3d}	D_{4h}	D_{6h}	O_h
p	C_1	C_2	D_2	D_3	D_4	D_6	O
m			C_{2v}	C_{3v}	C_{4v}	C_{6v}	
m		C_s			D_{2d}	D_{3h}	T_d
i				S_6	C_{4h}	C_{6h}	T_h
p				C_3	C_4	C_6	T
m					S_4	C_{3h}	

6.3 Asymmetric Domains

In crystallographic textures, besides the fundamental notion of "orientation" also the concept of "misorientation" is used. To each orientation of a crystallite corresponds a rotation leading from a laboratory (sample) Cartesian reference frame to the reference frame attached to the crystallite. The misorientation between two crystallites is the rotation leading from the reference frame of the first crystallite to the frame of the second one. From the formal viewpoint, an orientation can be treated as misorientation between the crystallite and the laboratory, and it is not necessary to consider it separately. Below, whenever a general orientation relationship is considered, the term "misorientation" is used.

Since crystal structures are symmetric, a number of rotations correspond to the misorientation, and the rotation parameters are not unique. To make the parameters unique, instead of the whole rotation space, only a part part of it is considered. The part is constructed in such a way that each physically distinct misorientation is represented there only once. Many different names were used to call that region, e.g. asymmetric domain, asymmetric unit, asymmetric region, symmetrically equivalent area, fundamental zone or Mackenzie cell.

In general, the symmetries of misoriented crystallites may differ. This occurs in multi–phase polycrystalline materials with different crystallographic symmetries of the phases. Analogous situation is encountered in the case of a symmetric crystal embedded in a symmetric sample.

Let two crystallites have symmetries described by point groups G_L and G_R. With S_L and S_R being elements of G_L and G_R, points O_L and O_R of the orientation space representing orientations of the crystallites are equivalent to $S_R O_R$ and $S_L O_L$, respectively. The misorientation between the crystallites (the rotation leading from O_R to O_L) is represented by the point in the rotation space given by

$$O = O_L O_R^{-1} . \tag{6.2}$$

The product $O_L O_R^{-1}$ is equivalent to $S_L O_L O_R^{-1} S_R^{-1}$. This means that

$$\text{the points} \quad O \quad \text{and} \quad S_L O S_R \quad \text{are equivalent.} \tag{6.3}$$

Thus, the direct product of G_L and G_R acts in the space of rotations and divides it into classes of equivalent points. The asymmetric domain constitutes a suitably chosen set of unique representatives of the classes.

Our goal is to determine shapes of such domains in some parameterizations. However, since the shapes tend to be complicated, the complete solution for all crystallographic symmetries will be described only in the case of Rodrigues parameterization. Before that, the domains in the space of Euler angles will be given for some selected symmetry pairs. Moreover, equivalent Miller indices in the so–called cubic–orthorhombic case (O, D_2) will be considered. Further on, the symbol (G_L, G_R) will stand for a particular pair of symmetries; the first element in the parentheses represents the symmetry acting on the left side, and the second one is the symmetry acting on the right side. If G_R is the trivial symmetry group C_1, instead of (G, C_1) only the symbol G will be used.

The case of the single–phase material, i.e., with all crystallites of the same kind, has two special features. First, because the crystallites are indistinguishable, the crystal coordinate systems must be attached to each of them in the same way. Therefore, there exists a natural and unique choice for the reference alignment (or reference null misorientation). On the other hand, when the crystallites are of different types, there are various possible settings of the crystal coordinate systems and the reference misorientation is not unique.

For example
In the case of misorientations between   hcp and bcc phases, bases of the crystal coordinate systems can be chosen in the standard way, i.e., for the null misorientation, the bcc (001) plane is parallel to hcp (0001). In some special cases, however, other choices would be interesting, e.g., to have bcc (110) plane parallel to hcp (0001) like in the so called Burgers' orientation relationship (see chapter 9). ⊠

The possibility to set coordinate systems in various ways means that for crystallites of different species the asymmetric domain is not "fundamental". Its shape depends on the choice of the reference misorientation. Therefore, particular results are valid only for given settings of the crystal coordinate systems.

The one used here (and referred to as the standard choice of the systems) is based on the table of stereograms of three dimensional crystallographic point groups printed in *International Tables for X–Ray Crystallography* (1952) with horizontal e_1 axis and vertical e_2 axis, both in the plane of the paper, and with e_3 perpendicular to that plane. (The first of two settings for the monoclinic system is taken.)

The second feature of single–phase case has a statistical nature. If two misoriented crystallites are indistinguishable, their roles can be exchanged; hence, values of e.g., misorientation distribution at mutually inverse rotations (O and O^{-1}) are the same. Therefore, also these rotations are considered to be equivalent and thus, the asymmetric domain is reduced by half.

□ *"Rotation function"*
We are primarily interested in distributions of crystallites but there are other applications of asymmetric domains. For example, the domains are essential for analysis of so called rotation functions. A rotation function is defined as the correlation of two mutually rotated Patterson functions (Rossmann & Blow, 1962, Tollin & Rossmann, 1966). The rotation functions are used in search for structurally identical or similar molecular sub–units within large molecules. Symmetries of the rotation functions are of the same kind as the symmetries of misorientation distributions. The equivalence between inverse rotations occurs if a rotation function is an auto–correlation of a Patterson function (Moss, 1985). ⊠

6.3.1 Pairs involving improper symmetry operations

The complete rotation space is composed of two separate components of proper and improper rotations. The relation (improper rotation) = (proper rotation) ∘ (a fixed improper rotation) provides a one–to–one correspondence between the components. Using it, a set of parameters can be ascribed to an improper rotation. Thus, in the general description involving improper rotations, a misorientation is determined by three parameters and a flag indicating to which of the two components the rotation belongs.

When improper symmetries are taken into account (i.e., the groups G_R and G_L are allowed to contain improper symmetry operations), crystals without such symmetries occur in two distinguishable enantiomorphic forms and the following two approaches are practicable:

1. ignore improper symmetries and consider the distinguishable forms as distinct phases with their own characteristics (e.g., textures),

2. have the domain of distributions extended on both, right- and left-handed forms, i.e., on both components of the complete rotation space. Presence of improper symmetry operations means that there are mutually equivalent points in the separate components.

The first approach seems to be favored by the texture community (cf. Bunge, 1982). Within this scheme, there are four[1] 'connected spaces' to be considered. In the case of the second approach, there are only two such spaces, i.e., two components of the rotation manifold; comparing to the first approach, the handedness of misoriented crystals is missing. This is not incorrect because misorientation, as it is defined, is not supposed to carry information about

---

[1] Two types of chirality of the first crystal times two chiralities for the second crystal.

original features of objects but only about their relative features. Knowing misorientation, it is impossible to guess the orientation of an object, and analogously, it is impossible to guess its handedness, unless, respectively, orientation and handedness of the other object are known.

In the important special case, when the first object is the laboratory coordinate system, the difference disappears because the handedness of this system is known *a priori*. Thus, in the first approach only two spaces remain, and in the second one, the orientation carries the information about the handedness of the second object.

In order to have a full solution to the problem of asymmetric domains in the sense of the first approach, it is enough to consider only proper rotations. At the outset, we use this approach, and all symmetry operations are assumed to be proper.

Assuming the second approach, the issue is more complex. However, for many symmetry pairs involving groups of the second kind, the forms of the domains can be deduced from the known domains obtained within the first approach and involving only proper rotations. Appropriate expressions will be listed below for the following separately considered five cases:

1. Both groups G_R and G_L are of the first kind. There are no equivalences between points in different components of the rotation manifold and thus, the asymmetric domain must consist of two parts; one part in each of the components. With the correspondence between proper and improper rotations established by the relation (improper rotation) = (proper rotation) ∘ (inversion), both components can be tessellated in the same way. The asymmetric domain can be obtained by choosing in each of the components the same region as the domain for (G_L, G_R).

   In all remaining cases, there are equivalences between points of the different components, and the asymmetric domain can be confined to the component of the identity.

2. Only one of the groups contains improper symmetry operations. Let it be G_L. The appropriate asymmetric domain, is the same as the domain for $(P(G_L), G_R)$, where $P(G)$ is the crystallographic subgroup of G consisting of proper rotations (Table 6.2).

3. Both groups G_R and G_L contain inversion. Inversion commutes with all rotations and thus, the configuration of crystals obtained by applying improper symmetry operations on both sides of their misorientation, can be obtained by applying proper operations of the symmetry groups. The asymmetric domain is the same as that for $(P(G_L), P(G_R))$.

4. Both groups G_R and G_L are of the second kind but only one of them, say G_L, contains inversion. It is possible to have equivalent rotations leading to distinguishable configurations; this is illustrated in Fig. 6.1. To simplify the description of the asymmetric domain, let $I(G)$ denote a group generated by generators of G and inversion. Moreover, let $Q(G) := P(I(G))$ (Table 6.2). The asymmetric domain in the space of proper rotations is the same as the domain of $(P(G_L), Q(G_R))$.

5. Both groups G_R and G_L are of the second kind but none of them contains inversion. Analogously to the previous point, the application of improper symmetry operations to both sides of proper rotation may lead to a configuration different than the one obtained by application of proper symme-

try operations. As for the asymmetric domains, they must be determined separately for particular pairs because there is no simple relation to the results involving only proper rotations.

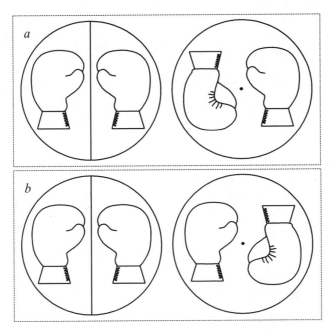

Fig. 6.1. The symmetries of the two objects are mirror (left object) and inversion (right object). Simultaneous application of the mirror and inversion symmetries is equivalent to the rotation by π about the axis perpendicular to the mirror plane (because inversion commutes with all other elements of the orthogonal group and *mirror ∘ inversion* is equal to that rotation). If right gloves are related by translation in the initial configuration (a), then this occurs for left gloves in the configuration (b). Thus, despite their equivalency, the two configurations are distinguishable.

6.3.2 Asymmetric domains in Euler space

The most common method, and in many cases the most convenient   one, is to determine the asymmetric domains in the Euler angles parameterization (Tollin, Main & Rossmann (1966), Pospiech (1972), Pospiech, Gnatek & Fichtner (1974), Rao, Jih & Hartsuck (1980), Zhao & Adams (1988)). Its main advantage is that for symmetries other than cubic, relations between equivalent sets of angles are linear.

A simple way to see how the domains are constructed is by an example. Let $O(\varphi_1, \phi, \varphi_2)$ be the rotation matrix corresponding to Euler angles $\varphi_1, \phi, \varphi_2$. Let us consider a crystal with symmetry described by the group D_4 with the principal axis along e_3, and the twofold axis along e_2. The symmetry operation of rotation about e_3 by $\pi/2$ is represented by $\mathrm{diag}(-1, -1, 1)$. For

Table 6.2. Groups $P(G)$ and $Q(G)$ corresponding to the crystallographic groups (G) of the second kind. The relation $P(D_{3h}) = D_3^{(2)}$, where $D_3^{(2)}$ denotes D_3 with principal axis along e_3 and twofold axis along e_2, appears in the first table because in the *International Tables for X–Ray Crystallography* (1952) the twofold axis of D_{3h} is fixed as vertical (e_2) whereas it is horizontal (e_1) for D_3.

For a group G of the first kind, there occurs $P(G) = Q(G) = G$.

G	C_s	S_4	C_{2v}	C_{3h}	C_{3v}	C_{4v}	D_{2d}	C_{6v}	D_{3h}	T_d
$P(G)$	C_1	C_2	C_2	C_3	C_3	C_4	D_2	C_6	$D_3^{(2)}$	T
$Q(G)$	C_2	C_4	D_2	C_6	D_3	D_4	D_4	D_6	D_6	O

G	C_i	C_{2h}	C_{3i}	C_{4h}	C_{6h}	D_{2h}	D_{3d}	D_{4h}	D_{6h}	T_h	O_h
$P(G) = Q(G)$	C_1	C_2	C_3	C_4	C_6	D_2	D_3	D_4	D_6	T	O

the symmetry acting on the left side, because $\operatorname{diag}(-1, -1, 1)O(\varphi_1, \phi, \varphi_2) = O(\varphi_1, \phi, \varphi_2 + \pi/2)$, the orientation $(\varphi_1, \phi, \varphi_2)$ is equivalent to $(\varphi_1, \phi, \varphi_2 + \pi/2)$. Since the symmetry operation can be applied numerous times, the orientation $(\varphi_1, \phi, \varphi_2)$ is actually equivalent to $(\varphi_1, \phi, \varphi_2 + k\pi/2)$, where k is an integer. Based on this equivalence, the range of φ_2 can be reduced from (2.28) to $0 \leq \varphi_2 < \pi/2$. Additionally, the second generating symmetry operation – the rotation about e_1 by π – is represented by the matrix $\operatorname{diag}(1, -1, -1)$. Since $\operatorname{diag}(1, -1, -1)O(\varphi_1, \phi, \varphi_2) = O(\pi + \varphi_1, \pi - \phi, \pi - \varphi_2)$, the orientation $(\varphi_1, \phi, \varphi_2)$ is equivqalent to $(\pi + \varphi_1, \pi - \phi, \pi - \varphi_2)$. Summarising, this allows us to take as the asymmetric domain the region limited by

$$0 \leq \varphi_1 < 2\pi , \quad 0 \leq \phi \leq \pi/2 , \quad 0 \leq \varphi_2 < \pi/2 .$$

There are no further reductions because the region is $1/8$ of the complete space and the order of the group D_4 is eight.

There may be also a symmetry operation acting on the right side; let it correspond to D_2 with the symmetry axes along e_i. Considerations similar to those given above show that the domain

$$0 \leq \varphi_1 < \pi/2 , \quad 0 \leq \phi \leq \pi/2 , \quad 0 \leq \varphi_2 < \pi/2 \qquad (6.4)$$

is asymmetric. Here and furhter, equivalences between boundary points of asymmetric domains are disregarded. To take them into account, additional rules would be necessary. But these points constitute a set of measure 0 and do not play any role in analysis of distributions, which are of our primary interest.

The above example corresponds to the so–called 'tetragonal – orthorhombic' case (D_4, D_2). It is related to the important and frequently used 'cubic – orthorhombic' case of textures in rolled metal sheets; the symmetries are the 'cubic' crystal symmetry and the 'orthorhombic' symmetry of a rolled sample. The threefold axis along $e_1 + e_2 + e_3$, which differs the cubic symmetry O form the tetragonal D_4, causes additional equivalences. These equivalences, however, cannot be expressed in a linear form. Therefore, results which in reality belong to the 'cubic – orthorhombic' case are usually presented in the 'tetragonal – orthorhombic' asymmetric domain (6.4). For the 'cubic – orthorhombic' symmetry, this domain contains symmetrically equivalent points, i.e., it is not asymmetric. See chapter 9 for some special orientations presented in that region.

Let us concentrate on non–cubic symmetries in general. Within the framework of proper rotations, there are only four cases to be considered (C_n, C_m), (D_n, C_m), (C_n, D_m) and (D_n, D_m). The generating equivalences and inequalities determining customary asymmetric domains are listed below. Operations on φ_1 and φ_2 are modulo 2π. The integers k_m take the values from 1 to m.

- (C_n, C_m)
  Equivalences:

$$(\varphi_1, \phi, \varphi_2) \leftrightarrow (\varphi_1 + 2k_m\pi/m, \phi, \varphi_2 + 2k_n\pi/n),$$

  Volume of the asymmetric domain: $1/(mn)$.
  Asymmetric domain: $0 \leq \varphi_1 < 2\pi/m$, $\quad 0 \leq \phi \leq \pi$, $\quad 0 \leq \varphi_2 < 2\pi/n$.
- (D_n, C_m)
  Equivalences:

$$(\varphi_1, \phi, \varphi_2) \leftrightarrow (\varphi_1 + 2k_m\pi/m, \phi, \varphi_2 + 2k_n\pi/n),$$
$$(\varphi_1, \phi, \varphi_2) \leftrightarrow (\pi + \varphi_1, \pi - \phi, \pi - \varphi_2).$$

  Volume of the asymmetric domain: $1/(2mn)$.
  Asymmetric domain: $0 \leq \varphi_1 < 2\pi/m$, $\quad 0 \leq \phi \leq \pi/2$, $\quad 0 \leq \varphi_2 < 2\pi/n$.
- (C_n, D_m)
  Equivalences:

$$(\varphi_1, \phi, \varphi_2) \leftrightarrow (\varphi_1 + 2k_m\pi/m, \phi, \varphi_2 + 2k_n\pi/n),$$
$$(\varphi_1, \phi, \varphi_2) \leftrightarrow (\pi - \varphi_1, \pi - \phi, \pi + \varphi_2).$$

  Volume of the asymmetric domain: $1/(2mn)$.
  Asymmetric domain: $0 \leq \varphi_1 < 2\pi/m$, $\quad 0 \leq \phi \leq \pi/2$, $\quad 0 \leq \varphi_2 < 2\pi/n$.
- (D_n, D_m)
  Equivalences:

$$(\varphi_1, \phi, \varphi_2) \leftrightarrow (\varphi_1 + 2k_m\pi/m, \phi, \varphi_2 + 2k_n\pi/n),$$
$$(\varphi_1, \phi, \varphi_2) \leftrightarrow (\pi + \varphi_1, \pi - \phi, \pi - \varphi_2),$$
$$(\varphi_1, \phi, \varphi_2) \leftrightarrow (\pi - \varphi_1, \pi - \phi, \pi + \varphi_2).$$

  Volume of the asymmetric domain: $1/(4mn)$.
  Asymmetric domain: $0 \leq \varphi_1 \leq \pi/m$, $\quad 0 \leq \phi \leq \pi/2$, $\quad 0 \leq \varphi_2 < 2\pi/n$.

6.3.3 Miller indices in the cubic-orthorhombic case

In metallurgy, crystallite orientations are frequently specified by Miller indices of the rolling plane and rolling direction. (See section 2.9.) Because of symmetry, some orientations given by different sets of indices are physically equivalent. For octahedral crystal symmetry, all symmetrically equivalent orientations can be created by applying the following rule: Values of indices are permuted simultaneously in both () and []. If the permutation is even, the number of '−' signs added to the original values of indices must be even (0 or

2). If the permutation is odd, the number of bars added to the original values of indices must be odd (1 or 3). The rule is obtained by using (6.3) and (2.62).

⊡ *Example*
Orientations equivalent to $(123)[63\bar{4}]$ are:

$$
\begin{array}{llll}
(1\,2\,3)[6\,3\,\bar{4}] & (2\,\bar{1}\,3)[3\,\bar{6}\,\bar{4}] & (2\,3\,1)[3\,\bar{4}\,6] & (\bar{1}\,2\,3)[\bar{6}\,3\,\bar{4}] \\
(3\,\bar{2}\,1)[\bar{4}\,\bar{3}\,6] & (\bar{1}\,3\,2)[\bar{6}\,\bar{4}\,3] & (3\,1\,2)[\bar{4}\,6\,3] & (\bar{2}\,1\,3)[\bar{3}\,6\,4] \\
(\bar{2}\,\bar{3}\,1)[\bar{3}\,4\,6] & (1\,\bar{3}\,2)[6\,\bar{4}\,3] & (\bar{2}\,3\,\bar{1})[\bar{3}\,\bar{4}\,6] & (3\,2\,\bar{1})[4\,3\,\bar{6}] \\
(\bar{3}\,2\,1)[4\,3\,6] & (\bar{3}\,1\,2)[4\,\bar{6}\,3] & (2\,\bar{3}\,\bar{1})[3\,4\,6] & (1\,3\,\bar{2})[6\,\bar{4}\,3] \\
(\bar{3}\,1\,\bar{2})[4\,6\,3] & (3\,\bar{1}\,2)[4\,\bar{6}\,3] & (\bar{3}\,\bar{2}\,\bar{1})[4\,3\,\bar{6}] & (\bar{1}\,\bar{3}\,2)[\bar{6}\,4\,3] \\
(2\,1\,\bar{3})[3\,6\,4] & (\bar{1}\,2\,\bar{3})[\bar{6}\,3\,4] & (1\,\bar{2}\,3)[6\,\bar{3}\,4] & (\bar{2}\,\bar{1}\,3)[\bar{3}\,\bar{6}\,4]
\end{array}
$$

.

☒

The materials to which this method of determining orientations is usually applied have the crystal symmetry O_h which involves inversion. In this case the plane $(h\,k\,l)$ is equivalent to $(\bar{h}\,\bar{k}\,\bar{l})$, and there is no distinction between $[u\,v\,w]$ and the direction $[\bar{u}\,\bar{v}\,\bar{w}]$. Despite the fact that only the crystal symmetry was used to get the equivalent sets of indices, the orientations $(h\,k\,l)[u\,v\,w]$, $(\bar{h}\,\bar{k}\,\bar{l})[u\,v\,w]$, $(h\,k\,l)[\bar{u}\,\bar{v}\,\bar{w}]$ and $(\bar{h}\,\bar{k}\,\bar{l})[\bar{u}\,\bar{v}\,\bar{w}]$ are not equivalent in an asymmetric sample. The orientations are equivalent if the sample has orthorhombic symmetry and (within reasonable approximation) this is the case of rolled materials. With cubic–orthorhombic symmetry, all simultaneous permutations of indices in () and [] with all sign combinations are possible, and the only constraint is that the specified direction lies in the specified plane. That is why it is common to omit the the bars over indices, i.e., we write {123} <634> instead of {123} <63$\bar{4}$>.

6.4 Asymmetric Domains in Rodrigues Space

Determination of asymmetric domains will be approached in a more systematic way for the Rodrigues parameterization. There are two reasons for that. First, for cubic symmetries the true asymmetric domains in Euler angles are awkward. Second, in the Rodrigues space, the domains can be constructed in such a way that they contain rotations with smallest rotation angles, and this property is helpful in analysis of misorientations. It will be used in chapter 7. An effort was made to cover solutions for all crystallographic point groups, and this required some technicalities. In order to get a general idea of the method, a cursory reading of this section is sufficient.

Let the distance between orientations represented by Gibbs vectors r_1 and r_2 be the smallest rotation angle (of rotations leading from r_1 to r_2), i.e., the quantity corresponding to (3.4). The distance is given by $2\arctan(|r_2 \circ (-r_1)|)$. In the Rodrigues space, points at the same distance to each of two given points constitute two *planes* (in the Euclidean sense). For our purposes, only planes equidistant to the zero Gibbs vector $0 = (0,0,0)$ and $r = \tan(\omega/2)l$ are needed; l is a unit vector and ω is in the range $(0, \pi]$. These planes consist of all points $\tan^{\pm 1}(\omega/4)l + y$, where y is an arbitrary vector perpendicular to l (cf. Frank, 1988, Morawiec, 1995).

| □ | *Set of points equidistant to two given points* |

Let $r = \tan(\omega/2)l$ $(r \neq 0)$ and, moreover, let x be a Gibbs vector representing an orientation with the same angular distance to r as to 0. The vectors representing 'the orientation differences' between r and x and between 0 and x are given by $(-x) \circ r$ and $(-x) \circ 0 = -x$, respectively. The equality of angular distances means that

$$((-x) \circ r) \cdot ((-x) \circ r) = (-x) \cdot (-x) .$$

This is the equation for x. It can be transformed to

$$(r \cdot x)^2 + 2(r \cdot x) - (r \cdot r) = 0 .$$

Its only solution has the form $x = \xi r + y$ where y is an arbitrary vector perpendicular to r and $\xi = (-1 \pm \sqrt{1 + r \cdot r})/(r \cdot r)$. Thus, the points at the same angular distance from 0 and from r lie on two planes, both perpendicular to r, and located on both sides of the origin at the (Euclidean) distances

$$\sqrt{(\xi r) \cdot (\xi r)} = \frac{\sqrt{1 + \tan^2(\omega/2)} \pm 1}{\tan(\omega/2)} = \tan^{\mp 1}(\omega/4) ,$$

where ω is in the range $(0, \pi)$. The conclusion is also valid for r at infinity as the limit for ω approaching π. ⊠

6.4.1 Pairs without non–trivial common symmetries

A symmetry operation present in both G_L and G_R is considered to be a common symmetry operation in specified coordinate systems if it has the same parameters when expressed explicitly in both systems. To give an example, let two crystals have symmetries described by the crystallographic point groups T and C_3. Each of these groups contains the symmetry with threefold rotation axis. If in both crystal coordinate systems this axis is along the same line (e.g., e_3 axis) this symmetry element is considered to be common. It is not a common element if the systems are chosen in the standard way with one of the twofold axes of T along e_3.

If the pair (G_L, G_R) has *no* common symmetry operations other than the identity (0 Gibbs vector), each misorientation is represented in the Rodrigues space by $K = \#G_L \#G_R$ different symmetrically equivalent points, where $\#G$ denotes the order of the group G. Those equivalent to 0 are given by the Gibbs vectors $s_L \circ 0 \circ s_R = s_L \circ s_R$, where s_L and s_R are Gibbs vectors representing elements of G_L and G_R, respectively. The points represented by $s_L \circ s_R$ will be described as distinguished points of the orientation space. The space can be divided into regions with boundary points located at the same angular distance from two nearest distinguished points. Points inside such region are closer to one of $s_L \circ s_R$ than to any other. The method of obtaining these regions is the same as the tessellation into the so called 'Voronoi polyhedra'. In the case of the Rodrigues parameters, regions' boundaries are planar, due to the fore–mentioned fact that points equidistant to two distinguished points lie on planes. The angular distance of a given point r to 0 is the same as the distance between $s_L \circ r \circ s_R$ and $s_L \circ s_R$. Thus, disregarding points at the boundaries, each region contains representatives of all classes of non-equivalent points and each class is represented only once in the region. Therefore, each of the regions can serve as the asymmetric domain. It is most convenient to choose the one surrounding 0; points of this region correspond to rotations with the smallest rotation angles.

Voronoi tessellation

Voronoi tessellation is the partitioning of a Euclidean space with a set of k different distinguished points into k convex polytopes such that each polytope contains all points that are closer to one of the distinguished points than to any other point of the set. This geometrical construction is sometimes called Dirichlet tessellation. Wigner–Seitz cell and Brillouin zone (in reciprocal space) used in physics are constructed in the same way. The idea of using Voronoi tessellation for determination of asymmetric domains in Euler angles parameterization was presented by Yeates (1993) in the context of rotation functions. $\boxtimes$

To determine the asymmetric domain for a particular symmetry, all distinguished points $s_L \circ s_R$ must be taken into consideration. Excluding $0 \circ 0$, they are of the form $l_{(i)} \tan(\omega_{(i)}/2)$, where $i = 1, 2, ..., (K-1)$, $l_{(i)}$ are unit vectors and $0 < \omega_{(i)} \le \pi$. The planes equidistant to the i-th point and to 0 are given by $(l_{(i)}; \tan^{\pm 1}(\omega_{(i)}/4))$, where the symbol $(l; d)$ denotes the plane perpendicular to the unit vector l at the distance d from 0, in the direction indicated by the vector. For $0 < \omega < \pi$ there occurs $\tan(\omega/4) < 1 < \tan^{-1}(\omega/4)$; moreover, $\tan^{\pm 1}(\pi/4) = 1$. Thus, the asymmetric domain composed of points located closer to 0 than to any other distinguished point is a common of all half–spaces determined by the conditions

$$l_{(i)} \cdot x \le \tan(\omega_{(i)}/4) \quad \text{if} \quad 0 < \omega_{(i)} < \pi ,$$
$$\pm l_{(i)} \cdot x \le 1 \quad \text{if} \quad \omega_{(i)} = \pi ,$$

where x indicates a point of the space.

Example: the pair (C_3, T)

Let us consider the pair (C_3, T) with crystal coordinate systems chosen in the standard way. There are no common symmetry elements other than the identity. Points $\{0, 0, \pm \tan(\pi/6)\}$ of the Rodrigues space represent symmetry operations of C_3. Because of them, the asymmetric domain is bounded by two planes $((0, 0, \pm 1); a)$, where $a = \tan(\pi/12) = 2 - 3^{1/2}$. Moreover, with the elements $s_L = \{0, 0, -\tan(\pi/6)\}$ of C_3 and $s_R = \{1, 1, 1\}$ of T, the distinguished point $s_L \circ s_R$ has coordinates $\{1, a, a\}$ and its distance from 0 is $b = (1 + 2a^2)^{1/2}$. Its presence leads to another bounding plane. It is given by $((\alpha, \beta, \beta); c)$, where $\alpha = 1/b$, $\beta = a/b$ and $c = \tan(\arctan(b)/2) = \beta(1+a)/(1-a)$. In the same way, a list of eight bounding planes is obtained; their parameters are $((\alpha, \pm\beta, \pm\beta); c)$, $((\beta, \pm\alpha, \mp\beta); c)$, $((-\alpha, \pm\beta, \mp\beta); c)$ and $((-\beta, \pm\alpha, \pm\beta); c)$, where either upper or lower signs are valid. The shape of the domain is shown in Fig. 6.2. $\boxtimes$

The following general property reduces the number of pairs to be determined: the asymmetric domain of the pair (G_R, G_L) can be obtained by taking inversion of the domain of (G_L, G_R) with respect the origin of the Cartesian system in the space of Rodrigues parameters.

Especially simple are the domains for the trivial symmetry C_1 on one side. Assuming that $G_R = C_1$, the asymmetric domain is a polyhedron filled with points which are closer to 0 than to any other of s_L ($\ne 0$) points. With the symbol h_n denoting $h_n := \tan(\pi/(2n))$, the domains for individual symmetries can be described in the following way:

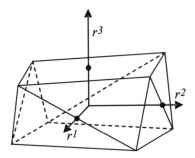

Fig. 6.2. Schematic drawing of the asymmetric domains for (C_3, T). The top and bottom faces are squares on parallel planes.

- *Cyclic symmetries*
  For $G_L = C_1$ the asymmetric domain overlaps the entire Rodrigues space. For C_n $(n \neq 1)$ the domain is bounded by two planes perpendicular to the n–fold axis, each at the distance h_n from 0.
- *Dihedral symmetries*
  The asymmetric domain for $G_L = D_n$ is a prism with $2n$-sided polygons (at the distance h_n from 0) as prism bases, and $2n$ square prism faces at the distance $h_2 = 1$. The bases are perpendicular to the n–fold axis and the faces are perpendicular to the twofold axes. See Fig. 6.3a.
- *Tetrahedral symmetry*
  The domain for the group T is the regular octahedron with faces at the distance $h_3 = \sqrt{3}/3$ from 0 and perpendicular to the threefold axes.
- *Octahedral symmetry*
  The asymmetric domain is a truncated cube with six octagonal faces at the distance $h_4 = \sqrt{2} - 1$ from the origin, and eight triangular faces at the distance $h_3 = \sqrt{3}/3$ (Handscomb, 1958). The octagonal faces are perpendicular to the fourfold symmetry axes and triangular faces are perpendicular to the threefold axes. See Fig. 6.4a.

$\boxed{\square}$ *Icosahedral symmetry*
  The asymmetric domain is the regular dodecahedron with faces at the distance h_5 from the origin. The faces are perpendicular to the fivefold symmetry axes. $\boxtimes$

6.4.2 Pairs with common symmetries

It is easy to notice that the approach applied to pairs without common symmetry elements must be modified when such elements are present. It follows directly from the fact that some of the products $s_L \circ s_R$ are no longer different. There are distinguished points overlapping each other and for them the planes of equidistant points cannot be constructed.

It is a natural step to consider the Voronoi tessellation of the Rodrigues space based on distinct distinguished points. With this tessellation, the Voronoi cell surrounding 0 will be called 'large cell'. It contains misorientations with the smallest rotation angles but there are still equivalent points

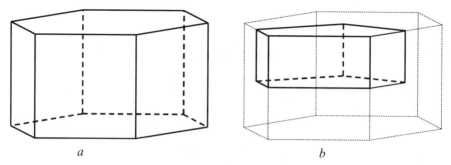

Fig. 6.3. Schematic representation of the asymmetric domains for (D_3, C_1) (a) and (D_3, D_3) (b).

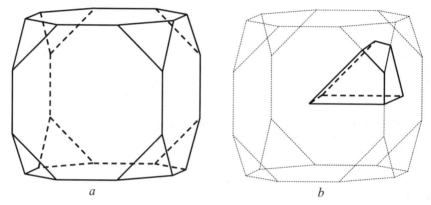

Fig. 6.4. Schematic representation of the asymmetric domains for (O, C_1) (a) and (O, O) (b).

(with equal rotation angles) inside it. This subsection contains details concerning equivalences within the large cell and clarifies the way of determining locations of additional surfaces cutting off the asymmetric domain from the cell.

Let the largest common divisor of integer numbers m and n be denoted by $\mathsf{LCD}(m, n)$. Moreover, the number $mn/\mathsf{LCD}(m, n)$ is briefly denoted by $\mathsf{F}(m, n)$. In order to investigate the role of the common symmetry elements let us concentrate on the case when rotations with the same rotation axis are applied on both sides of a Gibbs vector r, and the axis is m–fold on the right side and n–fold on the left one. In that case, the task of finding the asymmetric domain can be described by the following rules: There are two bounding planes perpendicular to the axis at the distance of $\tan(\pi/(2\mathsf{F}(m, n)))$ from 0. Moreover, the region between these planes can be radially divided (like a pie) into $\mathsf{LCD}(m, n)$ identical and equivalent parts, of which one can be selected as the asymmetric domain (Fig. 6.5). The latter division can be performed by introducing $\mathsf{LCD}(m, n)$ symmetrically distributed half–planes having the rotation axis as their common edge, with the dihedral angle between closest pairs equal to $2\pi/\mathsf{LCD}(m, n)$.

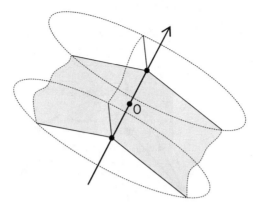

Fig. 6.5. Tessellation of the Rodrigues space caused by the presence of a common 3–fold rotation axis. The region between the parallel planes and the shaded surfaces may serve as the asymmetric domain.

Auxiliary construction

This is to justify the statements given above. The symmetry operations on the left and right sides have the same axis determined by the unit vector l. To have a nontrivial common symmetry operation, the numbers m and $n(< m)$ must have a nontrivial common divisor ($\mathsf{LCD}(m, n) \neq 1$). The points symmetrically equivalent to r are of the form $r' = s_L \circ r \circ s_R$, where $s_R = \tan(i_m \pi/m)l$ and $s_L = \tan(i_n \pi/n)l$, with $i_m = 0, ..., m - 1$ and $i_n = 0, ..., n - 1$. Let a plane perpendicular to l in the Rodrigues space be determined by all r satisfying $r \cdot l = \tan(\omega/2) = const$. Brief calculation shows that $r' \cdot l = \tan(\omega/2 + i_m \pi/m + i_n \pi/n)$. This means that the plane perpendicular to l and located at the distance of $\tan(\omega/2)$ from 0 is transformed onto a plane perpendicular to l and at the distance of $|\tan(\omega/2 + (i_m/m + i_n/n)\pi)|$. Due to the periodicity of tan, the number of overlapping planes (i.e., the number of planes with the same value of $r' \cdot l$) is equal to the number of different pairs (i_m, i_n) satisfying $(i_m/m + i_n/n)\pi = k\pi$ with $k = 0, 1$. This number, in turn, is equal to $\mathsf{LCD}(m, n)$. The total number of pairs (i_m, i_n) is mn. Thus, the number of distinct images of the initial plane is $\mathsf{F}(m, n)$.

There are $\mathsf{F}(m, n)$ distinct distinguished points $\tan(k\pi/\mathsf{F}(m, n))l$, $k = 0, ..., \mathsf{F}(m, n) - 1$, equivalent to 0. Two points nearest to 0 are $\pm \tan(\pi/\mathsf{F}(m, n))l$. They give rise to the two planes $(\pm l; \tan(\pi/(2\mathsf{F}(m, n))))$ bounding the zone of points close to 0. But there are still $\mathsf{LCD}(m, n)$ equivalent points inside that zone. To eliminate them, equivalences within the transformed plane must be considered. The image, r', is in the same plane as the initial point r, when $r' \cdot l = r \cdot l$. Let α be the angle between projections $h = r - \tan(\omega/2)l$ and $h' = r' - \tan(\omega/2)l$ of the vectors r and r' on the considered plane (Fig. 6.6). For r' satisfying $r' \cdot l = r \cdot l$ there occurs $r' \cdot r' = r \cdot r$ and thus, $h' \cdot h' = h \cdot h$. Hence, for r not collinear with the rotation axis, $\cos(\alpha) = (h' \cdot h)/(h \cdot h) = \cos(2i_m \pi/m)$, where i_m $(0 \leq i_m < m)$ satisfies $i_m n + i_n m = kmn$ for $i_n = 0, ..., n - 1$ and $k = 0, 1$. This means that $\alpha = 2\pi j/\mathsf{LCD}(m, n)$, $j = 0, ..., \mathsf{LCD}(m, n)$. Thus, all $\mathsf{LCD}(m, n)$ equivalent points on the considered plane are at the same distance from the axis l; they are distributed symmetrically around that axis and therefore, the plane can be divided radially into $\mathsf{LCD}(m, n)$ equivalent regions with angles between dividing half–lines equal to $2\pi/\mathsf{LCD}(m, n)$. ☒

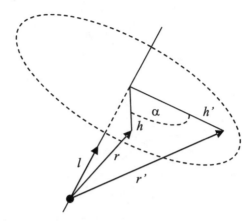

Fig. 6.6. Sketch illustratating quantities used for exploring equivalences within a plane perpendicular to the symmetry axis.

Positions of the dividing half–planes in the Rodrigues space are not unique but, if there is more than one axis corresponding to common symmetry operations, the half–planes must be properly arranged. It is *not* sufficient to establish proper dihedral angles and to satisfy the condition that each half–plane has the corresponding axis as its edge; the asymmetric domain must be built of such points r that application of all angle–preserving symmetry operations of the type $s \circ r \circ (-s)$ to the points of the domain, gives the large cell. All common symmetry operations constitute a group. It is a crystallographic subgroup of both G_L and G_R. Thus, the subgroup determines the positions of the half–planes. Knowing the subgroup, one can easily deduce which points of the large cell are equivalent and where the half–planes should be located.

The following notation allows us to make the description of the domains more concise. Let us assume that (G'_L, G'_R) is a pair without common symmetry operations. The expression

$$(G_L, G_R) \rightarrow (G'_L, G'_R)/G_C$$

will be used, when the asymmetric domain of (G'_L, G'_R) constitutes the large cell for (G_L, G_R), and the group G_C determines equivalences (due to (G_L, G_R)) between points of the large cell. In this section, where only proper rotations are considered, G_C is simply the group of common elements of G_L and G_R. In this case, the volume of the asymmetric domain of (G_L, G_R) is $\#G_C$ times smaller than the volume of the large cell. Once the asymmetric domain of (G'_L, G'_R) is known, based this notation, it is straightforward to deduce the form of the domain of the pair (G_L, G_R).

If one of the groups, say G_L, is a crystallographic subgroup of the other one (G_R), the coordinate systems can be arranged in such a way that all elements of the subgroup are common symmetry operations, and the large cell of (G_L, G_R) is nothing else but the asymmetric domain of the group $G_R \sim (G_R, C_1)$ alone. Thus, there occurs $(G_L, G_R) \rightarrow G_R/G_L$. In particular, if $G_R = G = G_L$, then $(G, G) \rightarrow G/G$. For the domains of (D_3, D_3) and (O, O) see Figs. 6.3b and 6.4b.

□	*Example*

In order to give an example for the case with one group being crystallographic subgroup of the other, let us consider the pair (C_3, T) once more. This time, the threefold axes of both, C_3 and T, are assumed to have the same direction in the state of reference alignment (non–standard arrangement). Therefore, there are three half–planes in the Rodrigues space with threefold axis as their common edge and dihedral angles of $2\pi/3$, and the appropriate domain for (C_3, T) is one third of the actual domain for tetrahedral symmetry. Obviously, this region is different than the one described in the example given in subsection 6.4.1. ⊠

6.4.3 Solutions for proper crystallographic symmetries

To solve the problem for proper crystallographic symmetries, all possible pairs of 11 crystallographic groups of the first kind must be considered. Most of these pairs satisfy subgroup–group relationship, i.e., one group is a crystallographic subgroup of the other. To take advantage of this fact, it would be necessary to choose the crystal coordinate systems in such way that the subgroup–group relation is geometrically satisfied,[2] and thus, in numerous cases, awkward crystal coordinate systems would have to be accepted. Therefore, further on, the standard choice of the systems will be assumed: for cyclic and dihedral symmetries the principal axis is along e_3, and the twofold axis of the latter is along e_1; for T and O two– and fourfold axes, respectively, are along axes of the Cartesian coordinate system. Particular pairs are considered below. Results, from which asymmetric domains can be deduced, are summarized in Table 6.3.

The form of the asymmetric domain for combinations involving only cyclic symmetries follows directly from the 'auxiliary construction'. The region contained between the planes $((0, 0, \pm 1); \tan(\pi/(2\mathsf{F}(m, n))))$ represents the large cell of (C_m, C_n). On the other hand, it also is the asymmetric domain of $C_{\mathsf{F}(m,n)}$. The common subgroup is $C_{\mathsf{LCD}(m,n)}$. Using the introduced notation, these statements can be expressed by

$$(C_m, C_n) \;\rightarrow\; C_{\mathsf{F}(m,n)}/C_{\mathsf{LCD}(m,n)} \; .$$

Also the pair (C_m, D_n) can easily be solved. Its large cell has the same form as the asymmetric domain of $D_{\mathsf{F}(m,n)}$. The group of common symmetry operations is $C_{\mathsf{LCD}(m,n)}$. The half–planes cutting the domain of $D_{\mathsf{F}(m,n)}$ have their edges on the principal axis. In the symbolic notation, the result has the form

$$(C_m, D_n) \;\rightarrow\; D_{\mathsf{F}(m,n)}/C_{\mathsf{LCD}(m,n)} \; .$$

The domain for the pair (D_m, D_n) can be obtained by using the solution for (C_m, D_n) and by noticing that the only difference is due to additional common twofold rotation axis along e_1. This leads to two complementary half–planes perpendicular to e_3. Formally, the result can expressed as

$$(D_m, D_n) \;\rightarrow\; D_{\mathsf{F}(m,n)}/D_{\mathsf{LCD}(m,n)} \; ,$$

---

[2] I.e., if the symmetries represented by Gibbs vectors s_L constitute a crystallographic subgroup of symmetries represented by s_R, then each of the vectors s_L is equal to one of the vectors s_R.

where D_1 represents the symmetry with the twofold axis along e_1. (See Fig. 6.3b.)

For each of the already considered pairs, the distinguished points are the same as those of a single group. Among remaining non–trivial pairs (i.e., these not satisfying the subgroup–group relationship), the same can be applied to (C_4, T) and (D_4, T). For both pairs, the large cell is the asymmetric domain of the octahedral group

$$(C_4, T) \rightarrow O/C_2 \quad \text{and} \quad (D_4, T) \rightarrow O/D_2 .$$

In order to obtain results for the remaining pairs, let us notice that in the standard arrangement the pairs (C_6, T), (D_3, T) and (D_6, T) differ from (C_3, T) by symmetry elements which are common symmetry elements. Simple analysis shows that

$$(C_6, T) \rightarrow (C_3, T)/C_2 , \quad (D_3, T) \rightarrow (C_3, T)/D_1 , \quad (D_6, T) \rightarrow (C_3, T)/D_2$$

and thus, solutions for these systems follow from the first example for (C_3, T). Finally, for the last four non–trivial pairs there occurs

$$(C_6, O) \rightarrow (C_3, O)/C_2 , \quad (D_3, O) \rightarrow (C_3, O)/D_1 \text{ and } (D_6, O) \rightarrow (C_3, O)/D_2.$$

Therefore, to complete the task for all pairs of crystallographic groups of the first kind, it remains to consider (C_3, O). The asymmetric domain for this case is described below. A drawing of the domain can be found in (Morawiec, 1997). The related system (D_6, O) is an imortant combination from the practical viewpoint due to the application to metals with 'cubic' and 'hexagonal' phases. It is called "hexagonal–cubic" in papers by Bonnet (1980) and Heinz and Neumann (1991). See the latter article for a stereo–pair of the domain of the system (O, D_6).

The pair (C_3, O)

It is straightforward to find the first six planes bounding the asymmetric domain. Because of threefold (of C_3) and fourfold (of O) rotation axes along e_3, there exist bounding planes $((0, 0, \pm 1); \tan(\pi/24))$. Moreover, fourfold rotation axes (of O) along e_1 and e_2 lead to $((\pm 1, 0, 0); \tan(\pi/8))$ and $((0, \pm 1, 0); \tan(\pi/8))$, respectively. But there are still symmetrically equivalent points in the described region. To eliminate them, eight additional planes are required. These are the same planes as those given in the first example for (C_3, T): $((\alpha, \pm\beta, \pm\beta); c)$, $((\beta, \pm\alpha, \mp\beta); c)$, $((-\alpha, \pm\beta, \mp\beta); c)$, $((-\beta, \pm\alpha, \pm\beta); c)$. (Either upper or lower signs are valid.) The symbols a, b, c, α and β denote the same quantities as in the case of (C_3, T). ☒

6.4.4 Solutions for groups with improper symmetry operations

The previous subsection, where only proper rotations were considered, provides the full solution to the problem of asymmetric domains in the sense of the first approach described in section 6.3.1. With improper symmetries involved, we have new Voronoi tessellations and new asymmetric domains. What considerably simplifies the task of determining the domains is that they can be described in a way similar to that used in the previous sections, i.e., by

Table 6.3. Asymmetric domains of the pairs (G_L, G_R) with groups G_L and G_R containing proper symmetry operations. The first column and the first row contain groups G_L and G_R, respectively. Entries of the table correspond to the right side of the arrow $\rightarrow$. The large cell of (G_L, G_R) is the same as the asymmetric domain of the pair (group) given on the left side of the slash '/'. The group on the other side of the slash determines the symmetry of sets of equivalent points within the large cell. The entries on the diagonal, i.e., for $G_R = G = G_L$, are given by G/G.

	O	T	D_6	D_4	D_3	C_6	D_2	C_4	C_3
C_2	O/C_2	T/C_2	D_6/C_2	D_4/C_2	D_6	C_6/C_2	D_2/C_2	C_4/C_2	C_6
C_3	(C_3,O)	(C_3,T)	D_6/C_3	D_{12}	D_3/C_3	C_6/C_3	D_6	C_{12}	
C_4	O/C_4	O/C_2	D_{12}/C_2	D_4/C_4	D_{12}	C_{12}/C_2	D_4/C_2		
D_2	O/D_2	T/D_2	D_6/D_2	D_4/D_2	D_6/D_1	D_6/C_2			
C_6	$(C_3,O)/C_2$	$(C_3,T)/C_2$	D_6/C_6	D_{12}/C_2	D_6/C_3				
D_3	$(C_3,O)/D_1$	$(C_3,T)/D_1$	D_6/D_3	D_{12}/D_1					
D_4	O/D_4	O/D_2	D_{12}/D_2						
D_6	$(C_3,O)/D_2$	$(C_3,T)/D_2$							
T	O/T								

specifying $(G'_L, G'_R)/G_C$ with G'_L, G'_R and G_C being groups of the first kind. The group on the right side of the slash (G_C) still determines equivalences between points with the same rotation angles but it does not represent the group of common symmetry operations any more.

In four of the five cases considered in section 6.3.1, the shapes of domains follow directly from the solution for groups with improper symmetry operations. The fifth case must be considered separately because there is no simple relation to the results of the previous sections. Let both groups G_R and G_L be of the second kind and none of them contain inversion. As before, the subgroup diagram can be used to reduce the number of cases which should be considered. If the group G_L is crystallographic subgroup of G_R then there exists a setting in which all elements of G_L are common symmetry operations. In such case, all distinguished points in the first component correspond to elements of $P(G_R)$, i.e., the asymmetric domain of the pair is part of the domain of $P(G_R)$. More precisely, it is given by $P(G_R)/Q(G_L)$. Also here, however, that approach has the disadvantage of enforcing awkward settings for crystal coordinate systems. Te results corresponding to the standard settings of the coordinate systems (*International Tables for X–Ray Crystallography*, 1952) are given in Table 6.4.

Table 6.4. Asymmetric domains of the pairs (G_L, G_R) for groups G_L and G_R of the second kind without inversion. The diagonal entries of the table for $G_R = G = G_L$ are given by $P(G)/Q(G)$, e.g., for (T_d, T_d) one has $P(T_d)/Q(T_d) = T/O$. The symbol $C_2^{(i)}$ denotes C_2 with rotation axis along e_i, and $D_n^{(i)}$ denotes D_n with principal axis along e_3 and twofold axis along e_i. The twofold axis e_4 of $D_n^{(4)}$ is the bisector of the angle between e_1 and e_2.

	T_d	D_{3h}	C_{6v}	D_{2d}	C_{4v}	C_{3v}	C_{3h}	C_{2v}	S_4
C_s	O	$D_3^{(2)}/C_2$	D_6	D_4	D_4	$D_3^{(2)}$	C_3/C_2	D_2	C_4
S_4	T/C_4	D_{12}	$D_6^{(4)}/C_2$	D_2/C_4	D_4/C_2	$D_6^{(4)}$	C_{12}	$D_2^{(4)}/C_2$	
C_{2v}	O/C_2	$D_6/C_2^{(1)}$	C_6/D_2	D_4/C_2	C_4/D_2	$C_6/C_2^{(1)}$	D_6		
C_{3h}	(C_3,O)	$D_3^{(2)}/C_6$	D_6/C_3	D_{12}	D_{12}	$D_3^{(2)}/C_3$			
C_{3v}	(C_3,O)	$D_3^{(2)}/D_3$	C_6/D_3	D_{12}	$C_{12}/C_2^{(1)}$				
C_{4v}	$O/D_2^{(4)}$	$D_{12}/C_2^{(1)}$	C_{12}/D_2	$D_4/D_2^{(4)}$					
D_{2d}	T/D_4	$D_{12}/C_2^{(2)}$	D_{12}/C_2						
C_{6v}	$(C_3,O)/C_2$	D_6/D_3							
D_{3h}	$(C_3,O)/C_2^{(2)}$								

7

Misorientation Angle and Axis Distributions

T HE misorientation angle ω between two objects is defined as the smallest of rotation angles among equivalent rotations relating two given orientations of the objects[1]. It is the simplest characteristic of the difference between orientations of two crystallites in a polycrystalline material. Measured distributions of misorientation angles are compared with a special function: the distribution of the misorientation angles obtained in the case when orientations of crystallites are random. This particular distribution is a reference for distributions of misorientation angles of grains in real materials with non–random textures and orientation correlations (Mackenzie, 1958, Grimmer, 1979a,b). In relation to misorientation angles, there is also a question about the distributions of corresponding axes for randomly oriented crystallites. Such a function for octahedral symmetry was calculated by Mackenzie (1964). For other cases see, e.g., Morawiec (1996a, 1997). Again, the distributions of rotation axes corresponding to randomly oriented crystallites constitute a reference for distributions occurring in real materials. The misorientation angle and the corresponding axis depend on symmetries of the crystallites (or objects in general). Therefore, also the distributions are influenced by the symmetries.

The situation when the misoriented crystallites are identical is most typical. However, the misorientation angle and axis can also be determined for crystallites of different types and possibly different symmetries, as in multiphase materials. In that case, because of the possibility to set coordinate systems in arbitrary ways, the form of misorientation angle distributions depends on the choice of the reference misorientation. Here, the same settings as in the previous chapter are used.

Following conventional approach, it is appropriate to consider the rotation angle only for proper rotations. Thus, in order to calculate the distributions of misorientation angles, only parts of the asymmetric domains contained inside the first component of the rotation manifold are needed.

As an additional remark closing this introduction, it is worth noting that distributions of an arbitrary rotation parameters can be determined. The in-

---

[1] Sometimes, equivalent orientation differences are referred to as misorientations, and those with the smallest rotation angle – as "disorientations". In that case, the smallest rotation angle is called "disorientation" angle.

terest in the rotation angle and axis is caused by the fact that they are perceived as the most natural parameters. Moreover, the rotation angle is directly related to the distance between orientations and the axis complements this set of parameters. Besides that, however, no fundamental meaning can be ascribed to the misorientation angle and axis distributions.

7.1 Misorientation Angle Distributions

The simplest way to determine the sought distributions is by the use of Rodrigues parameters r^k ($k = 1, 2, 3$), because in this case the asymmetric domains containing rotations with the smallest rotation angles are known. With a domain of this kind, a misorientation angle distribution at ω can be obtained by integrating the density of random orientations (3.10) over the surface $\omega = \text{const}$ within the domain. The density of random orientations in the Rodrigues space is related to the invariant volume of the space. Using spherical coordinates with the radius $\mathfrak{r} = \sqrt{r^k r^k}$, the expression for the invariant volume $dV(r) = (\pi(1 + r^k r^k))^{-2} dr^1 dr^2 dr^3$ can be re-written as

$$dV = \frac{1}{\pi^2} \frac{\mathfrak{r}^2}{(1 + \mathfrak{r}^2)^2} d\mathfrak{r} \, d_2 n \tag{7.1}$$

where $d_2 n$ is the infinitesimal element on the unit sphere.

With the volume of the whole rotation space set to be 1, the volume of the asymmetric domain equals $1/K$, where K is an integer. Taking into account the normalization coefficient, the distribution $\mathfrak{p}$ of $\mathfrak{r}$ is given by

$$\mathfrak{p}(\mathfrak{r}) = \frac{K}{\pi^2} \frac{\mathfrak{r}^2}{(1 + \mathfrak{r}^2)^2} \, \chi(\mathfrak{r}) \,, \quad \text{where} \quad \chi(\mathfrak{r}) := \int_{\Omega(\mathfrak{r})} d_2 n$$

and $\Omega(\mathfrak{r})$ denotes the solid angle based on this part of the sphere of the radius $\mathfrak{r}$ which is comprised inside the asymmetric domain (Fig. 7.1).

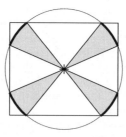

Fig. 7.1. Two dimensional sketch showing the solid angle which contributes to the integral $\chi(\mathfrak{r})$ (shaded area). The asymetric domain is represented by the rectangle.

The angle ω of misorientation of r with respect to the reference point $0 = (0, 0, 0)$ is given by $\tan(\omega/2) = \mathfrak{r}$. Since $\mathfrak{r}$ is unambiguously related to ω, the expression for $\mathfrak{p}(\mathfrak{r})$ is the basis for calculating the sought misorientation angle distribution $p(\omega)$; based on $p(\omega) = \mathfrak{p}(\tan(\omega/2))\,(d\mathfrak{r}/d\omega)(\omega)$, we have

$$p(\omega) = \frac{K}{2\pi^2}\sin^2(\omega/2)\,\chi(\tan(\omega/2))\,.$$

Summarizing, in order to calculate the misorientation angle distribution for a given symmetry the function $\chi(\mathfrak{r})$ is needed, and it can be obtained from the geometry of the asymmetric domain.

The misorientation angle distribution is actually determined by the large cell of a given pair of point groups. If the asymmetric domain of (G'_L, G'_R) is the large cell for the pair (G_L, G_R), then the misorientation angle distribution for (G_L, G_R) are the same as for (G'_L, G'_R). These statements follow from the fact that the geometric figure of the large cell has (at least) the point symmetry of the common subgroup. Transformation of the outer planes bounding the asymmetric domain (i.e, these which are not through 0) by elements of the subgroup gives the planes bounding the large cell. But these outer planes of the domain constitute the only factor influencing the sought misorientation angle distribution. Therefore, to obtain the distribution, the large cell can be used. In order to have proper normalization of the distribution, the normalizing factor must correspond to the volume of the large cell. Thus, K is equal to $\#G'_L\#G'_R$ or

$$K = \#G_L\#G_R/\#G_C\,.$$

This guarantees that the formulae for p in the cases without common symmetry operations are also correct when such operations are present.

It is easy to see that the misorientation angle distributions for (G_L, G_R) and (G_R, G_L) are identical. The pairs (G, G), (G, C_1) have the same large cell, and thus the same misorientation angle distributions. The list of large cells of crystallographic point groups in the standard setting is limited to the asymmetric domains of C_m, D_m, T, O, (C_3, T) and (C_3, O). The functions for all these cases are listed below. The expression $\tan(\pi/(2m))$ is abbreviated by h_m. Following Mackenzie's convention (Mackenzie, 1958), the numerical values of angles are given in degrees.

Cyclic groups: For C_1 the asymmetric domain overlaps with whole Rodrigues space and $\chi(\mathfrak{r}) = 4\pi$. Hence, we get the well known rotation angle distribution $(p(\omega) = (2/\pi)\sin^2(\omega/2))$ for the whole rotation space.

For C_m $(m \neq 1)$, the asymmetric domain is bounded by two parallel planes, each at the distance h_m from 0. Therefore, when $\mathfrak{r} \leq h_m$ the whole sphere is inside the asymmetric domain and $\chi(\mathfrak{r}) = 4\pi$. For $\mathfrak{r} > h_m$ two spherical caps, both with angular radii $\alpha_1(\mathfrak{r}) = \arccos(h_m/\mathfrak{r})$, are outside the asymmetric domain. The solid angle based on a spherical cap with angular radius of α is given by $2\pi(1 - \cos\alpha) =: S_1(\alpha)$. Thus, for $\mathfrak{r} > h_m$ two solid angles, each given by $S_1(\alpha_1(\mathfrak{r}))$, have to be subtracted from the full solid angle, and

$$\chi(\mathfrak{r}) = 4\pi - 2S_1(\alpha_1(\mathfrak{r}))\,.$$

The maximum misorientation angle is $180°$. The density functions for $m = 1, 2, 3, 4$ and 6 are shown in Fig. 7.2. Mean values are listed in Table 7.1.

☐☐ *Limiting case*
    By calculating the limit of $K\chi(\mathfrak{r})$ for $m \to \infty$, one obtains $K\chi(\mathfrak{r}) \to 2\pi^2/\mathfrak{r}$, and this allows us to get the distribution for objects with the axial symmetry C_∞.
☒

Table 7.1. Mean value for the misorientation angle distributions for some C_m groups.

Symmetry	$\overline{\omega}$
C_1	126.4756°
C_2	102.2958°
C_3	96.1595°
C_4	93.6981°
C_6	91.7618°

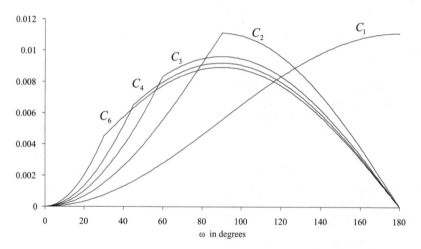

Fig. 7.2. Misorientation angle distributions for selected cyclic symmetries.

Dihedral groups: The asymmetric domain for D_m is a prism with $2m$-side polygons (at the distance h_m from 0) as prism bases, and $2m$ square prism faces at the distance $h_2 = 1$. The distance of the edges equals $\sqrt{1 + h_m^2}$, and the distance of the vertices is $\sqrt{1 + 2h_m^2}$. Thus, for $\mathfrak{r} \leq h_m$ the function $\chi(\mathfrak{r})$ is constant and equal to 4π. When $h_m < \mathfrak{r} \leq 1$, only spherical caps outside the basal planes have to be taken into account. In this case

$$\chi(\mathfrak{r}) = 4\pi - 2S_1(\alpha_1(\mathfrak{r}))$$

where $\alpha_1(\mathfrak{r}) = \arccos(h_m/\mathfrak{r})$. For $1 < \mathfrak{r} \leq \sqrt{1 + h_m^2}$, there are additional $2m$ spherical caps outside the prism faces and

$$\chi(\mathfrak{r}) = 4\pi - 2S_1(\alpha_1(\mathfrak{r})) - 2mS_1(\alpha_2(\mathfrak{r})) \ ,$$

where $\alpha_2(\mathfrak{r}) = \arccos(1/\mathfrak{r})$. Finally, when $\sqrt{1+h_m^2} < \mathfrak{r} \le \sqrt{1+2h_m^2}$, the sphere reaches edges of the prism and the caps outside partially overlap. Let $S_2(\alpha,\beta;\gamma)$ be the solid angle based on the common of two spherical caps of the angular radii α and β with centers at the angular distance γ. This function can be written in the form

$$S_2(\alpha,\beta;\gamma) = 2\left(\pi - C(\alpha,\beta,\gamma) - \cos\alpha\ C(\gamma,\alpha,\beta) - \cos\beta\ C(\beta,\gamma,\alpha)\right)$$

where $C(\alpha,\beta,\gamma) := \arccos\left((\cos\gamma - \cos\alpha\cos\beta)/(\sin\alpha\sin\beta)\right)$. Hence,

$$\chi(\mathfrak{r}) = 4\pi - 2S_1(\alpha_1) - 2mS_1(\alpha_2) + 4mS_2(\alpha_1,\alpha_2;\delta_1) + 2mS_2(\alpha_2,\alpha_2;\delta_2) \ ,$$

where $\delta_1 = \pi/2$ and $\delta_2 = 2\pi/(2m)$. The maximum misorientation angle and the mean for D_m, $m = 2, ..., 6$ are listed in Table 7.2 . See also Fig. 7.3.

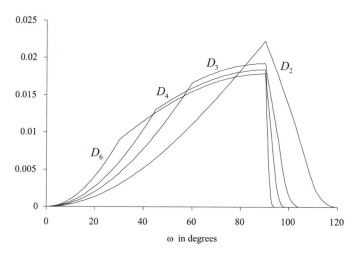

Fig. 7.3. Misorientation angle distributions for selected dihedral symmetries.

The common of two spherical caps

We are looking for the expression for the common part of two solid angles based on spherical caps. Let the angular radii of the caps be equal to α and β and the angular distance between centers of the caps be γ (Fig. 7.4).

Let us consider a spherical triangle with sides equal to α, β and γ, and the opposite vertex angles denoted by A, B and C, respectively. All of them are assumed not to exceed $\pi/2$. From the cosine theorem of spherical trigonometry, we have the expression for the angle $C = C(\alpha,\beta,\gamma)$

$$C(\alpha,\beta,\gamma) = \arccos\left(\frac{\cos\gamma - \cos\alpha\cos\beta}{\sin\alpha\sin\beta}\right)$$

Angles A and B can be obtained by simultaneous cyclic replacements: $A \to B \to C \to A$ and $\alpha \to \beta \to \gamma \to \alpha$, i.e., $A = C(\beta,\gamma,\alpha)$ and $B = C(\gamma,\alpha,\beta)$. Using the excess of the triangle ABC given by $E(\triangle ABC) = A + B + C - \pi$, we find out that the common part $S_2(\alpha, \beta; \gamma)$ of two solid angles equals to

$$\frac{A}{\pi} S_1(\beta) + \frac{B}{\pi} S_1(\alpha) - 2E(\triangle ABC) .$$

Thus, the sought function is

$$S_2(\alpha, \beta; \gamma) = 2 \left(\pi - C(\alpha, \beta, \gamma) - \cos\alpha \; C(\gamma, \alpha, \beta) - \cos\beta \; C(\beta, \gamma, \alpha) \right) .$$

⊠

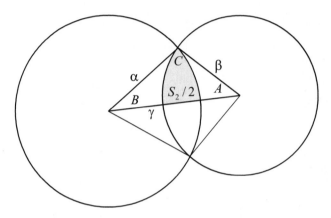

Fig. 7.4. Schematic diagram for the derivation of S_2.

Limiting case
It is worth noting that the limit of $K\chi(\mathfrak{r})$ for $m \to \infty$ is $K\chi(\mathfrak{r}) \to 4\pi^2/\mathfrak{r}$. This allows us to get the distribution corresponding to D_∞. ⊠

Table 7.2. Maximum misorientation angle and mean for some D_m groups.

Symmetry	Max. angle	$\bar{\omega}$
D_2	$120.0000°$	$75.1563°$
D_3	$104.4775°$	$66.6291°$
D_4	$98.4211°$	$63.0123°$
D_6	$93.8410°$	$60.0726°$

Tetrahedral group: The asymmetric domain is a regular octahedron with faces at the distance $\mathfrak{r}_1 = h_3 = \sqrt{3}/3$. The distances of its edges and vertices are $\mathfrak{r}_2 = \sqrt{2}/2$ and $\mathfrak{r}_3 = 1$, respectively. Proceeding like before, one gets three ranges with different behavior of $\chi(\mathfrak{r})$. With $\mathfrak{r}_0 = 0$, $\alpha_1(\mathfrak{r}) = \arccos(\mathfrak{r}_1/\mathfrak{r})$ and $\delta = \arccos(1/3)$ the formula for χ takes the form

$$\chi(\mathfrak{r}) = \sum_{i=1}^{j} \chi_i(\mathfrak{r}) \quad \text{for} \quad \mathfrak{r}_{j-1} \leq \mathfrak{r} < \mathfrak{r}_j \quad , \tag{7.2}$$

where $j = 1, 2, 3$ and

$$\chi_1(\mathfrak{r}) = 4\pi ,$$
$$\chi_2(\mathfrak{r}) = -8S_1(\alpha_1(\mathfrak{r})) ,$$
$$\chi_3(\mathfrak{r}) = 12S_2(\alpha_1(\mathfrak{r}), \alpha_1(\mathfrak{r}); \delta) .$$

The maximum misorientation angle equals $90°$. The mean misorientation angle is $\overline{\omega} = 51.4693°$. (See Fig. 7.5.)

Octahedral group: The asymmetric domain is a truncated cube with six octagonal faces at the distance $\mathfrak{r}_1 = h_4 = \sqrt{2} - 1$ from the origin, and eight triangular faces at the distance $\mathfrak{r}_2 = h_3 = \sqrt{3}/3$. The distance of all edges is $\mathfrak{r}_3 = \sqrt{2}h_4 = 2 - \sqrt{2}$ and the distance of vertices equals $\mathfrak{r}_4 = h_4\sqrt{5 - 2\sqrt{2}} = \sqrt{23 - 16\sqrt{2}}$. With $\mathfrak{r}_0 = 0$ and

$$\alpha_1(\mathfrak{r}) = \arccos((\mathfrak{r}_1/\mathfrak{r}), \quad \alpha_2(\mathfrak{r}) = \arccos(\mathfrak{r}_2/\mathfrak{r}), \quad \delta_1 = \pi/2 \text{ and } \delta_2 = \arccos(\sqrt{3}/3)$$

the expression for χ is given by Eq.(7.2), where $j = 1, ..., 4$ and

$$\chi_1(\mathfrak{r}) = 4\pi ,$$
$$\chi_2(\mathfrak{r}) = -6S_1(\alpha_1(\mathfrak{r})) ,$$
$$\chi_3(\mathfrak{r}) = -8S_1(\alpha_2(\mathfrak{r})) ,$$
$$\chi_4(\mathfrak{r}) = 12S_2(\alpha_1(\mathfrak{r}), \alpha_1(\mathfrak{r}); \delta_1) + 24S_2(\alpha_1(\mathfrak{r}), \alpha_2(\mathfrak{r}); \delta_2)$$

(Mackenzie, 1958 and Handscomb, 1958). The maximum and the mean misorientation angles are $62.7994°$ and $40.7358°$, respectively.

Icosahedral group
The asymmetric domain is a regular dodecahedron with faces at the distance $\mathfrak{r}_1 = h_5 = \tan(\pi/10)$. The distances of the edges and vertices are equal to

$$\mathfrak{r}_2 = \tan(\pi/10)\sin(\pi/5)\sec(\pi/3) \quad \text{and} \quad \mathfrak{r}_3 = \tan(\pi/10)\tan(\pi/5)\tan(\pi/3) ,$$

respectively. With $\mathfrak{r}_0 = 0$, $\alpha_1(\mathfrak{r}) = \arccos(\mathfrak{r}_1/\mathfrak{r})$ and $\delta = \arctan(2)$, the function $\chi = \chi(\mathfrak{r})$ is given by the sum (7.2), where $j=1, 2, 3$ and

$$\chi_1(\mathfrak{r}) = 4\pi ,$$
$$\chi_2(\mathfrak{r}) = -12S_1(\alpha_1(\mathfrak{r})) ,$$
$$\chi_3(\mathfrak{r}) = 30S_2(\alpha_1(\mathfrak{r}), \alpha_1(\mathfrak{r}); \delta) .$$

The maximum misorientation angle is $44.4775°$, and the mean misorientation angle equals $\overline{\omega} = 29.5127°$. ☒

The pair (C_3, T): The meaning of the symbols a, c, α and β is the same as in section 6.4.1. The points of interest on the abscissa are of the form $2\arctan(\mathfrak{r}_i)$ $(i = 1, .., 5)$, where $\mathfrak{r}_1$ and $\mathfrak{r}_2$ are distances (from 0) of the faces of the domain,

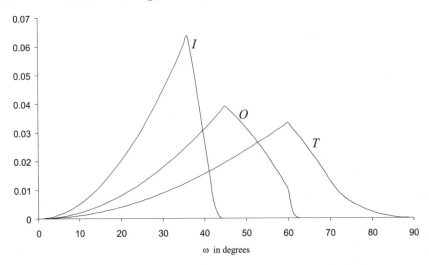

Fig. 7.5. Misorientation angle distributions for T, O and I.

$\mathfrak{r}_3$ and $\mathfrak{r}_4$ are distances of the edges, and $\mathfrak{r}_5$ is the distance of all vertices. They are given by

$$\mathfrak{r}_1 = a \ , \quad \mathfrak{r}_2 = c \ , \quad \mathfrak{r}_3 = 3^{1/2} a \ , \quad \mathfrak{r}_4 = a^{1/2} \ , \quad \mathfrak{r}_5 = 5^{1/2} a \ .$$

Angles between normals to the three types of neighboring planes are

$$\delta_1 = \arccos(\beta) \ , \quad \delta_2 = \arccos(\beta(2\alpha - \beta)) \ , \quad \delta_3 = \arccos(\alpha^2 - 2\beta^2) \ .$$

Let α_i $(i = 1, 2)$ be $\mathfrak{r}$–dependent quantities defined by $\alpha_i(\mathfrak{r}) = \arccos(\mathfrak{r}_i/\mathfrak{r})$. With $\mathfrak{r}_0 = 0$, the function χ is given by Eq.(7.2), where $j = 1, ..., 5$ and

$$\begin{aligned}
\chi_1(\mathfrak{r}) &= 4\pi \ , \\
\chi_2(\mathfrak{r}) &= -2S_1(\alpha_1(\mathfrak{r})) \ , \\
\chi_3(\mathfrak{r}) &= -8S_1(\alpha_2(\mathfrak{r})) \ , \\
\chi_4(\mathfrak{r}) &= 8S_2(\alpha_1(\mathfrak{r}), \alpha_2(\mathfrak{r}); \delta_1) + 4S_2(\alpha_2(\mathfrak{r}), \alpha_2(\mathfrak{r}); \delta_3) \ , \\
\chi_5(\mathfrak{r}) &= 4S_2(\alpha_2(\mathfrak{r}), \alpha_2(\mathfrak{r}); \delta_2) \ .
\end{aligned}$$

The distribution $(\pi/180)\, p(\omega\pi/180)$, is displayed in Fig. 7.6. The corresponding maximum misorientation angle is $61.8561°$. The mean misorientation angle is $36.4058°$.

The pair (C_3, O)**:**   The meaning of the symbols a, c, α and β is the same as in section 6.4.1 and as for the case of (C_3, T) above. The essential points on the abscissa are $2\arctan(\mathfrak{r}_i)$ $(i = 1, ..., 7)$, where $\mathfrak{r}_1$, $\mathfrak{r}_2$ and $\mathfrak{r}_3$ are distances (from 0) of the faces of the domain, $\mathfrak{r}_4$ and $\mathfrak{r}_6$ are distances of its edges, and $\mathfrak{r}_5$ an $\mathfrak{r}_7$ are distances of vertices. The radii $\mathfrak{r}_i$ are given by

$$\mathfrak{r}_1 = \tan(\pi/24) \ , \quad \mathfrak{r}_2 = \tan(\pi/8) \ , \quad \mathfrak{r}_3 = c \ , \quad \mathfrak{r}_4 = (\mathfrak{r}_1^2 + \mathfrak{r}_2^2)^{1/2} \ ,$$

$$\mathfrak{r}_5 = (\mathfrak{r}_1^2 + \mathfrak{r}_2^2 + \mathfrak{r}_1^2 \mathfrak{r}_2^2)^{1/2} , \quad \mathfrak{r}_6 = a^{1/2} , \quad \mathfrak{r}_7 = (a + \mathfrak{r}_1^2 + a\,\mathfrak{r}_1^2)^{1/2} .$$

Angles between normals to the four types of neighboring planes are δ_1, δ_2 (the same as for (C_3, T)), $\delta_3 = \arccos(\alpha)$ and $\delta_4 = \pi/2$. As before, α_i ($i = 1, 2, 3$) depend on $\mathfrak{r}$ and are given by $\alpha_i(\mathfrak{r}) = \arccos(\mathfrak{r}_i/\mathfrak{r})$. With $\mathfrak{r}_0 = 0$, the formula for χ has the form (7.2), where $j = 1, ..., 7$ and

$$\chi_1(\mathfrak{r}) = 4\pi ,$$
$$\chi_2(\mathfrak{r}) = -2S_1(\alpha_1(\mathfrak{r})) ,$$
$$\chi_3(\mathfrak{r}) = -4S_1(\alpha_2(\mathfrak{r})) ,$$
$$\chi_4(\mathfrak{r}) = -8S_1(\alpha_3(\mathfrak{r})) ,$$
$$\chi_5(\mathfrak{r}) = 8S_2(\alpha_1(\mathfrak{r}), \alpha_2(\mathfrak{r}); \delta_4) + 8S_2(\alpha_2(\mathfrak{r}), \alpha_3(\mathfrak{r}); \delta_3) + 8S_2(\alpha_1(\mathfrak{r}), \alpha_3(\mathfrak{r}); \delta_1) ,$$
$$\chi_6(\mathfrak{r}) = -8S_3(\alpha_1(\mathfrak{r}), \alpha_2(\mathfrak{r}), \alpha_3(\mathfrak{r}); \delta_3, \delta_1, \delta_4) ,$$
$$\chi_7(\mathfrak{r}) = 4S_2(\alpha_3(\mathfrak{r}), \alpha_3(\mathfrak{r}); \delta_2) .$$

Here $S_3(\rho_1, \rho_2, \rho_3; \xi_1, \xi_2, \xi_3)$ is the solid angle based on a common of three spherical caps with angular radii ρ_i ($i = 1, 2, 3$); the symbol ξ_k ($k = 1, 2, 3$) denotes the angular distance between the centers of the caps with radii ρ_i and ρ_j, where $i \neq j \neq k \neq i$. (For example, ξ_1 is the angular distance between the centers of caps with radii ρ_2 and ρ_3.) The function can be expressed as

$$
\begin{aligned}
S_3(\rho_1, \rho_2, \rho_3; \xi_1, \xi_2, \xi_3) = &-\pi \\
&+ (S_2(\rho_1, \rho_2; \xi_3) + S_2(\rho_2, \rho_3; \xi_1) + S_2(\rho_3, \rho_1; \xi_2))/2 \\
&+ \cos(\rho_1)C(\xi_2, \xi_3; \xi_1) + \cos(\rho_2)C(\xi_3, \xi_1; \xi_2) \\
&+ \cos(\rho_3)C(\xi_1, \xi_2; \xi_3) .
\end{aligned}
$$

The graph of the distribution is shown in Fig. 7.6. The maximum misorientation angle and the mean misorientation angle are 56.6003° and 33.2220°, respectively.

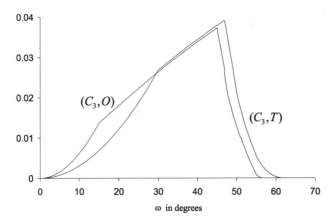

Fig. 7.6. Misorientation angle distributions for (C_3, T) and (C_3, O).

Solid angle based on the common of three spherical caps
    The solid angle S_3 based on the common of three spherical caps can be expressed as

$$S_3 = Q + \sum_{i=1}^{3} (R_i - P_i) \ ,$$

where

— Q is the solid angle corresponding to the hexagon $(A_1 B_3 A_2 B_1 A_3 B_2)$ with vertices at the centers A_i of the caps and the outer cross-points B_i of the caps' edges,

— P_i is the solid angle based on the part of i-th cap cut out by two radii of the cap; e.g., for the first cap the radii are $A_1 B_2$ and $A_1 B_3$; the part encompasses the common of the three caps; if ψ_i is internal vertex angle of the hexagon at the center of i-th cap, then $P_i = (\psi_i/(2\pi))S_1(\rho_i)$,

— R_i is the common of caps with radii ρ_j and ρ_k, where $i \neq j \neq k \neq i$ $(i, j, k = 1, 2, 3)$ and thus, $R_i = S_2(\rho_j, \rho_k; \xi_i)$. (See Fig. 7.7.)

Q is equal to the sum of solid angles based on the four spherical triangles $(A_1 B_3 A_2)$, $(A_2 B_1 A_3)$, $(A_3 B_2 A_1)$ and $(A_1 A_2 A_3)$, of which the hexagon is composed. For each spherical triangle, the solid angle is equal to the sum of its internal vertex angles minus π. The vertex angles, in turn, can be expressed through the side angles ρ_i and ξ_i using the cosine theorem of spherical geometry. Also ψ_i is the sum of three such angles with vertex at the center of i-th cap. The final result is reduced to the formula for $S_3(\rho_1, \rho_2, \rho_3; \xi_1, \xi_2, \xi_3)$ given above. $\boxtimes$

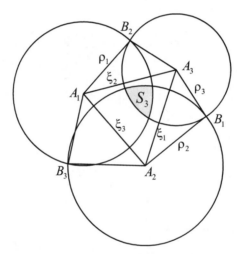

Fig. 7.7. Schematic diagram for the derivation of S_3.

7.2 Distributions of Rotation Axes

Since each axis is determined by a unit vector n, the distribution of axes q can be given as a distribution of such vectors $(q = q(n))$, i.e., as a normalized function on the unit sphere. In Rodrigues parameterization the invariant volume (and thus orientation density for the random orientation distribution) at the point $r = \mathfrak{r} n$ $(\mathfrak{r} \geq 0)$ is given by Eq.(7.1). In order to calculate the distributions of rotation axes for the rotations with the smallest rotation angles it is enough to integrate over these $\mathfrak{r}$ which represent points inside asymmetric domains of section 6.4. After taking into account the normalization coefficient, which is equal to K,

$$q(n) = \frac{K}{\pi^2} \int_0^{\lambda(n)} \frac{\mathfrak{r}^2}{(1+\mathfrak{r}^2)^2} \, d\mathfrak{r} \ ,$$

where $\lambda(n)$ is the Euclidean distance of the boundary of the asymmetric domain in direction n from the point 0. In the case of no symmetry, the range of integration extends to infinity and thus $q(n) = 1/(4\pi)$. For non-trivial symmetries, after integrating,

$$q(n) = \frac{K}{2\pi^2} \left(\arctan(\lambda(n)) - \frac{\lambda(n)}{1+(\lambda(n))^2} \right) \ ,$$

i.e., all we need to get the explicit expression for the axis distribution is the function λ. If $(l; d)$ is the domain bounding plane such that $n \cdot l > 0$, then $\lambda(n) = d/(n \cdot l)$. To be more specific, particular symmetries must be considered.

In the case without common symmetry operations, the distributions cover the sphere. If such operations are present the domain of the axis distribution is restricted to a part of the sphere ('standard triangle') but the function can be extended to the complete sphere by applying the common symmetry operations. In that case the distribution is actually determined by the large cell. As was already said, the list of large cells of crystallographic point groups in the standard setting consists of the asymmetric domains of C_m, D_m, T, O, (C_3, T) and (C_3, O) and they are considered below. Axis distributions for all remaining cases can be deduced from these results. The distributions for C_3, D_3, T, O are shown in Fig. 7.8.

Cyclic symmetries: For C_m, $(m \neq 1)$ the asymmetric domain is bounded by the planes $(\pm l; \tan(\pi/(2m)))$, where l is the unit vector of the m-fold symmetry axis. For $n \cdot l \neq 0$

$$\lambda(n) = \frac{\tan(\pi/(2m))}{\mid n \cdot l \mid} \ ,$$

while for $n \cdot l = 0$ the distance $\lambda(n)$ is infinite and thus $q(n) = m/(4\pi)$.

Dihedral symmetries: The asymmetric domain for D_m bounded by the planes $(\pm l; \tan(\pi/(2m)))$ and $(\pm l_{(i)}; 1)$, where l is the unit vector of the principal axis and $l_{(i)}$ are the unit vectors of two-fold axes $(i = 1, ..., m)$. Thus, $\lambda(n)$ is given by the formula

$$\lambda(n) = \min \left\{ \frac{\tan(\pi/(2m))}{\mid n \cdot l \mid} \ , \ \frac{1}{\max_i \{\mid n \cdot l_{(i)} \mid\}} \right\} \ .$$

Tetrahedral symmetry: The asymmetric domain for group T is the regular octahedron bounded by the planes $(\pm l_{(i)}; \sqrt{3}/3)$, where $l_{(i)}$ are the unit vectors of three-fold axes. Hence,

$$\lambda(n) = \frac{\sqrt{3}/3}{\max_i\{|\,n \cdot l_{(i)}\,|\}} \ .$$

Octahedral symmetry: The asymmetric domain is bounded by the planes $(\pm l_{(i)}; \sqrt{2} - 1)$ and $(\pm l'_{(j)}; \sqrt{3}/3)$, where $l_{(i)}$ are the unit vectors of four-fold axes and $l'_{(j)}$ are the unit vectors of three-fold axes. Therefore, $\lambda(n)$ is given by (Mackenzie, 1964)

$$\lambda(n) = \min\left\{ \frac{\sqrt{2} - 1}{\max_i\{|\,n \cdot l_{(i)}\,|\}} \ , \ \frac{\sqrt{3}/3}{\max_j\{|\,n \cdot l'_{(j)}\,|\}} \right\} \ .$$

Icosahedral symmetry
The asymmetric domain is a regular dodecahedron bounded by the planes $(\pm l_{(i)}; \delta)$, where $l_{(i)}$ are the unit vectors of five-fold axes, $\delta = \sqrt{(\sigma_1 - \sigma_2)/(\sigma_1 + \sigma_2)}$, $\sigma_1 = 2\sqrt{\tau}$, $\sigma_2 = \sqrt{1 + 2\tau}$ and τ is the golden section ratio $\tau = (1 + \sqrt{5})/2$. Thus,

$$\lambda(n) = \min_i\left\{ \frac{\delta}{|\,n \cdot l_{(i)}\,|} \right\} \ .$$

⊠

The expressions listed above are not restricted to any specific coordinate systems. The next two cases are given for the standard settings of coordinate systems.

The pair (C_3, T): Let (n_1, n_2, n_3) be components of a unit vector n determining a rotation axis. The function λ is given by

$$\lambda(n) = \min\left\{\mathfrak{r}_1/|\,n_3\,|, \mathfrak{r}_2/d_2\right\} \ ,$$

where
$$d_2 = \max\{\alpha n_1 \pm \beta n_2 \pm \beta n_3, \beta n_1 \pm \alpha n_2 \mp \beta n_3,$$
$$-\alpha n_1 \pm \beta n_2 \mp \beta n_3, -\beta n_1 \pm \alpha n_2 \pm \beta n_3\} \ ,$$

and the meaning of the symbols $\mathfrak{r}_1$, $\mathfrak{r}_1$, α and β is given in the description of misorientation angle distribution for (C_3, T) in section 7.1. In each of the four arguments of max, either upper or lower signs are valid.

The pair (C_3, O): The appropriate function λ has the form

$$\lambda(n) = \min\left\{\mathfrak{r}_1/|\,n_3\,|, \mathfrak{r}_2/\max\{|\,n_1\,|, |\,n_2\,|\}, \mathfrak{r}_3/d_2\right\} \ ,$$

where d_2 was already given above for (C_3, T) and the meaning of $\mathfrak{r}_1$, $\mathfrak{r}_2$ and $\mathfrak{r}_3$ is the same as in the description of misorientation angle distribution for (C_3, O) in section 7.1.

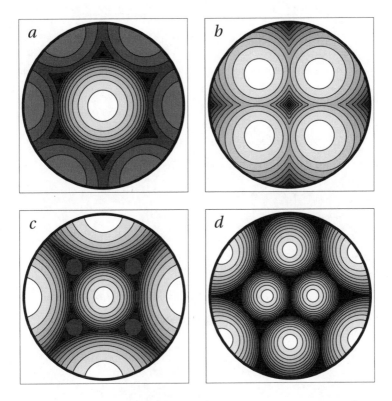

Fig. 7.8. Stereographic projection of the misorientation axis distributions for D_3 (*a*), T (*b*), O (*c*) and I (*d*). Axes for r_1 and r_2 of the coordinate frame are horizontal and vertical, respectively. The displayed isolines are equidistant in the range of the distribution ($isoline(i) = min + (i + .5)(max - min)/10$, $i = 0, 1, ..., 9$).

8

Crystalline Interfaces and Symmetry

THE analysis of intercrystalline boundaries constitutes an important area in the investigation of polycrystalline materials. This chapter contains a very brief account on parameter equivalences caused by crystal symmetries.

There is a basic dichotomous classification of boundaries into homophase and heterophase types (see, e.g., Sutton and Balluffi, 1995). The homophase (grain) boundaries separate grains of the same phase but with different orientations. The heterophase boundaries are inter–phase boundaries. A full description of a boundary requires parameters which would determine microscopic misfits between neighboring structures. However, experimental access to microscopic boundary features is difficult and a more crude description based on the so–called macroscopic boundary parameters is frequently used. The macroscopic parameters are: three numbers for the misorientation of grains, two for the local inclination of the boundary and one discrete parameter for the change of handedness. The continuous space in which boundaries are determined consists of rotations and unit vectors locally normal to the interfaces. The former may be identified with the manifold of orthogonal matrices $O(3)$, and the latter – via Gauss mapping – with the unit sphere S^2. Thus, the domain of boundaries is $O(3) \times S^2$.

8.1 Symmetrically Equivalent Boundaries

Because of crystal symmetries, the boundary parameters are not unique, i.e., a number of different sets of parameters represent the same geometrical arrangement at the boundary.[1] The domain occupied by all potential boundaries can be reduced to an asymmetric domain. Our aim is to investigate the parameter equivalency for both homophase and heterophase boundaries.

A boundary will be specified by an orthogonal matrix O representing the misorientation and a unit vector n normal to the boundary. It is assumed that a Cartesian coordinate system is attached to the crystal structure, the

---

[1] An interface and crystals which it separates, altogether treated as an autonomous object, may exhibit some symmetries. This issue, however, will not be considered here.

components of n are given in that system, and O relates the systems of two neighboring crystallites. To be precise, let us consider two neighboring crystallites labeled 1 and 2. All quantities related to the first crystal will have index 1 ascribed; similarly, 2 will indicate quantities related to the second crystal. Let the orientations of the crystallites with respect to an external coordinate system be given by orthogonal matrices O_1 and O_2. The boundary between the *first* grain and the *second* one is specified by the misorientation $O = O_1 O_2^T$ (see Eq.(6.2)) and the normal to the boundary n_1 (by convention) directed towards the second grain, with coordinates specified in the coordinate system of the first crystallite. The boundary is determined by the pair (O, n_1). The coordinates of the normal to the boundary in the Cartesian coordinate system of the second grain are given by $O^T n_1$. Thus, the grain boundary between the *second* grain and the *first* one has the form (O^T, n_2) with the normals n_1 and n_2 related by

$$n_2 = -O^T n_1 . \tag{8.1}$$

Due to crystal symmetries, different points of the space $O(3) \times S^2$ may represent geometrically identical interfaces. The presence of the symmetries leads to an equivalence between such points. The main question we are going to answer is: what are the relations between equivalent points.

Let n be a vector perpendicular to a crystallographic plane with Miller indices $(h \, k \, l) = h$ (Eq.2.63). Vector n^{eq} perpendicular a plane h^{eq} symmetrically equivalent to h is related to n via

$$n^{eq} = S \, n , \tag{8.2}$$

where S from $O(3)$ is the orthogonal part of a space group operation.

Explanation

The issue is a bit subtle because the plane may be polar, i.e., it may have distinguishable sides. To prove (8.2) let us consider non-coplanar vectors v_1, v_2, v_3 and a vector v_0; the endpoints of the first three vectors determine the crystallographic plane h, and the fourth vector v_0 indicates the sense of the normal. One can write $n = p \, \text{sign}(p \cdot (v_0 - v_1))$, where $p = (v_2 - v_1) \times (v_3 - v_1) = v_1 \times v_2 + v_2 \times v_3 + v_3 \times v_1$. An element $(S \mid t)$ of the crystal space group (t – translation) transforms v_μ ($\mu = 0, ..., 3$) into $v_\mu^{eq} = (S \mid t) v_\mu = S v_\mu + t$. Vector n^{eq} normal to the plane h^{eq} is determined by $v_0^{eq}, v_1^{eq}, v_2^{eq}, v_3^{eq}$. There occurs $(v_0^{eq} - v_1^{eq}) = S (v_0 - v_1)$ and $p^{eq} = v_1^{eq} \times v_2^{eq} + v_2^{eq} \times v_3^{eq} + v_3^{eq} \times v_1^{eq} = \det(S) \, S \, p$, and hence, $n^{eq} = p^{eq} \, \text{sign}(p^{eq} \cdot (v_0^{eq} - v_1^{eq})) = (\det(S))^2 \, S \, p \, \text{sign}(p \cdot (v_0 - v_1)) = S \, n$. ⊠

Let S_1 and S_2 be point symmetries of crystals 1 and 2, respectively. Thus, the orientation O_1 is equivalent to $S_1 O_1$. Analogously, O_2 is equivalent to $S_2 O_2$. Hence, the misorientations O and $S_1 O S_2^T$ are equivalent; cf. Eq.(6.3). The question is, when the misorientations $S_1 O S_2^T$ and planes n_1^{eq} and n_2^{eq} equivalent to n_1 and n_2, respectively, represent a boundary indistinguishable from the original one. From (8.2), the vectors n_1^{eq} and n_2^{eq} should satisfy $n_1^{eq} = S_1 n_1$ and $n_2^{eq} = S_2 n_2$. We want to know which of the crystal symmetries leave the relations between equivalent vectors and equivalent misorientations consistent. Application of a crystal symmetry operation, say S_1, may be seen as a change of the reference frame attached to the first crystal, and such a change has a simultaneous effect on both the misorientation and the normal.

The matrix O and the vector n_1 are transformed to $S_1 O$ and $S_1 n_1$, respectively. What matters here is that the element S_1 preceding n_1 cannot be decoupled from S_1 preceding O. Similarily, S_2 transforms O and n_2 to $O S_2^T$ and $S_2 n_2$. This means that

$$(O, n_1) \quad \text{is equivalent to} \quad (S_1 O S_2^T, S_1 n_1) \tag{8.3}$$

for all elements S_1 and S_2 of the point groups. In agreement with (8.1), $(S_1 O S_2^T)^T S_1 n_1 = S_2 O^T n_1 = -S_2 n_2$.

If inversion $\mathfrak{I}$ on both sides is involved, there is also equivalency between (O, n_1) and $(S_1 O S_2^T, -S_1 n_1)$; in particular,

$$(O, n_1) \quad \text{is equivalent to} \quad (O, -n_1) . \tag{8.4}$$

But this case is already included in (8.3) with $S_1 = \mathfrak{I} = S_2$. Finally, in the case of homophase boundaries when the crystallites are indistinguishable, we may consider

$$(O, n_1) \quad \text{to be equivalent to} \quad (O^T, n_2) = (O^T, -O^T n_1) .$$

This equivalence is of different nature than (8.3) because it is not based on crystal symmetry.

Analogously to the case of misorientations, an asymmetric domain is defined as a region in the parameter space in which each geometrically distinct boundary is represented only once, i.e., which contains exactly one representative of each equivalence class. The shapes of asymmetric domains follow directly from the domains for the misorientation distribution. The later are known for all combinations of crystallographic symmetries; see section 6.4. This allows asymmetric domains for all types of homo– and heterophase boundaries to be constructed; for instance, one can take the product of the asymmetric domain for misorientations (determined by the point groups of the crystals) and the unit sphere (or a hemisphere if (8.4) occurs).

It is easier to remember and to apply the symmetry relations (8.3) by using the following notation: let 4×4 matrices $\mathcal{B}$, $\mathcal{S}_1$ and $\mathcal{S}_2$ be defined by

$$\mathcal{B} = \begin{bmatrix} 0 & n_2^T \\ n_1 & O \end{bmatrix} , \quad \mathcal{S}_1 = \begin{bmatrix} 1 & 0 \\ 0 & S_1 \end{bmatrix} , \quad \mathcal{S}_2 = \begin{bmatrix} 1 & 0 \\ 0 & S_2 \end{bmatrix} ,$$

i.e., the first matrix (with $n_2 = -O^T n_1$) determines the boundary (O, n_1), and the other two correspond to the crystal symmetries. The matrix corresponding to the boundary (O^T, n_2) has the form $\mathcal{B}^T$. The equivalency (8.3) between (O, n_1) and $(S_1 O S_2^T, S_1 n_1)$ can be expressed as equivalence between the matrices $\mathcal{B}$ and $\mathcal{S}_1 \mathcal{B} \mathcal{S}_2^T$. These relations are similar to the equivalence relations for misorientations. It is noteworthy that $\det(\mathcal{B}) = \det(O)$, and $\mathrm{tr}(\mathcal{B}) = \mathrm{tr}(O)$, i.e., misorientation angles are related to traces of the 4×4 boundary matrices.

8.2 Boundary Distributions

One of the aspects of boundary investigation is to analyse their occurence statistically for all possible grain misorientations and boundary inclinations.

The frequency of given boundary types is represented by a distribution over the macroscopic boundary parameters. Such a distribution would play a role similar to that of the orientation distribution in crystallographic texture analysis. In order to define a boundary distribution, a measure in the parameter space must be specified. The measure, however, is not unique, and thus the boundary distribution depends on its choice. The situation is different than in the case of the orientation distribution which is determined on the special orthogonal group with unique invariant volume element. In the more general case, the expression for a volume element follows from the metric structure determining the distance between points of the manifold, i.e., representing the degree of closeness between boundaries. It was already mentioned in section 3.4 that with parameters x^i ($i = 1, ..., 5$) and metric tensor g, the volume element $\mathrm{d}\mathcal{V}$ is given by $\mathrm{d}\mathcal{V} = \sqrt{|\det(g)|}\,\mathrm{d}x^1\,\mathrm{d}x^2...\,\mathrm{d}x^5$.

Once the volume element $\mathrm{d}\mathcal{V}$ is given, the boundary distribution, say f, can be defined. Its value at the point $\mathcal{B}$ times an infinitesimal volume $\mathrm{d}\mathcal{V}$ centered at that point is equal to the ratio of the area $\mathrm{d}A$ of boundaries with parameters in $\mathrm{d}\mathcal{V}$ to the complete area of boundaries A

$$\mathrm{d}A/A = f(\mathcal{B})\,\mathrm{d}\mathcal{V} \ .$$

The element $\mathrm{d}\mathcal{V}$ is expected to satisfy $\int \mathrm{d}\mathcal{V} = 1$, and the distribution corresponding to $f = 1$ is considered to be random.

In the case of $SO(3) \times S^2$, the natural choice of the metric is the product metric of the unique invariant metric of $SO(3)$ and the unique canonical metric on S^2 inherited from the Euclidean space. Let us initially concentrate on finite distances. For two points (boundaries) (O, n) and (O', n'), the finite distance on $SO(3)$ can be given by $\chi_\bullet(O, O')$ defined by Eq.(3.5), and the finite distance on S^2, say $\chi_\circ$, is related to the angle γ between the vectors n and n'. It is assumed that $\chi_\circ^2(n, n') = (n - n')^2 = 2(1 - n \cdot n') = 2(1 - \cos(\gamma))$. (For small ϵ there occurs $2(1 - \cos(\epsilon)) = \epsilon^2 + \mathcal{O}(4)$, and therefore local metric properties given directly by ω and γ are the same as for $\chi_\bullet(O, O')$ and $\chi_\circ(n, n')$, respectively.)

The finite distance $\underline{\chi}$ between (O, n) and (O', n') could be defined by

$$\underline{\chi}^2((O, n), (O', n')) = \chi_\bullet^2(O, O') + \chi_\circ^2(n, n') \ .$$

This approach, however, has certain deficiency. Let us consider boundaries between grains A_1 and A_2, and between B_1 and B_2. We would like the distance between the boundaries A_1/A_2 and B_1/B_2 to be equal to the distance between A_2/A_1 and B_2/B_1. Does it occur for χ? The answer is 'no' but the distance χ can be modified to make it satisfy that condition. The condition is satisfied for χ defined as

$$\chi^2((O, n_1), (O', n_1')) = \chi_\bullet^2(O, O') + (\chi_\circ^2(n_1, n_1') + \chi_\circ^2(n_2, n_2'))/2 \ , \qquad (8.5)$$

with the normals related through Eq.(8.1), i.e., $n_2 = -O^T n_1$ and $n_2' = -O'^T n_1'$.

Boundaries can be compared by checking their distance χ. Similar boundaries are close in the space and thus their distance is small. Let us also mention that with $\mathcal{B}$ and $\mathcal{B}'$ corresponding to (O, n) and (O', n'), respectively, the distance (8.5) can be expressed in the form

$$\chi^2(\mathcal{B}, \mathcal{B}') = \|\mathcal{B} - \mathcal{B}'\|^2/2 \; .$$

Finally, it has to be taken into account that symmetries affect the finite metric properties of the space; in the symmetric case, the finite distance between two boundaries is given by the smallest of all values of χ for all representatives of the classes to which the boundaries belong.

Singularity of the distance (8.5)
Due to the part involving inclinations, the distance (8.5) is generally non–zero even if $O = I = O'$. For a perfect single crystal with two imaginary planes having different inclinations, there is no structural difference between such 'null' boundaries. With identical boundaries having a non–zero distance, (8.5) has a kind of 'singularity' at these points. The problem does not exist in the case of, say, antiphase boundaries or if the boundaries separate different phases. ⊠

Now, the local metric properties and the metric tensor can be obtained from the finite distance. To give an example, let us concentrate on a specific choice of parameters. For the unit sphere S^2, the spherical coordinates (α, β) with $\alpha \in [0, \pi]$ and $\beta \in [0, 2\pi)$ are defined in such a way that the Cartesian coordinates of n are $n^1 = \sin \alpha \cos \beta$, $n^2 = \sin \alpha \sin \beta$ and $n^3 = \cos \alpha$. In these coordinates the metric on S^2 can be expressed in the well known form $d\chi_\circ^2 = d\alpha^2 + \sin^2 \alpha \, d\beta^2$. (It is obtained by determining square of the distance of points with parameters (α, β) and $(\alpha + d\alpha, \beta + d\beta)$.) Thus, the metric tensor is diagonal with 1 and $\sin^2 \alpha$ on the diagonal. With total volume normalized to 1, the volume element $dV_\circ$ on the sphere is $dV_\circ = (4\pi)^{-1} \sin \alpha \, d\alpha \, d\beta$.

As for the metric on $SO(3)$, with Euler angles $\varphi_1, \phi, \varphi_2$ as coordinates, the invariant metric has the form $d\chi_\bullet^2 = d\varphi_1^2 + 2\cos(\phi)d\varphi_1 d\varphi_2 + d\varphi_2^2 + d\phi^2$. The invariant volume element is given by $dV_\bullet = (8\pi^2)^{-1} \sin(\phi) \, d\varphi_1 \, d\phi \, d\varphi_2$ (section 3.1).

For this particular choice of macroscopic boundary parameters, the product metric on $SO(3) \times S^2$ is $d\chi^2 = d\alpha^2 + \sin^2 \alpha \, d\beta^2 + d\varphi_1^2 + 2\cos(\phi)d\varphi_1 d\varphi_2 + d\varphi_2^2 + d\phi^2$, and the corresponding volume element has the form

$$dV = (32\pi^3)^{-1} \sin(\alpha) \sin(\phi) \, d\alpha \, d\beta \, d\varphi_1 \, d\phi \, d\varphi_2 \; .$$

Again, the volume is normalized to 1.

Going back to the boundary distribution, by symmetry it must take equal values for equivalent boundary parameters, i.e.,

$$f(\mathcal{B}) = f(\mathcal{S}_1 \mathcal{B} \mathcal{S}_2^T) \; , \tag{8.6}$$

where $\mathcal{S}_1$ and $\mathcal{S}_2$ represent symmetry operations. Other scalar functions given on boundaries may exhibit these symmetries. For example, if the Gibbs free energy of grain boundaries is assumed to be dependent only on macroscopic boundary parameters, it will satisfy relations analogous to (8.6).

9

Crystallographic Textures

CRYSTALLOGRAPHIC texture analysis deals with orientations of crystallites – building blocks of polycrystalline materials. Many common materials are polycrystalline. Among them are metals, ceramics, rocks, crystalline polymers (both, synthetic and natural) and composites of various kinds. Textures of metals have been getting most attention and considerable amount of knowledge has been collected on them. Currently, the texture analysis of metals is the best developed branch of crystallographic texture research.

Initially, the term 'crystallographic texture' meant 'crystallite orientation distribution' but the sense of the term evolved and become broader. Nowadays, besides the global approach, orientations are explored also locally. The so–called microtexture inherently involves analysis of misorientations. Also here, after the first section devoted to orientation distributions in the 'cubic–orthorhombic case', there are two sections on crystallites' relative orientations. Only some selected facts from the vast field of global textures will be given. Much more information can be found e.g., in the book edited by Kocks, Tomé and Wenk (1998). In the sections on misorientations, the accent is on explanation of essential concepts and constructions.

The uniform (random) distribution of crystallites – the one with the same probability of occurrence for all crystalline orientations – is exceptional. Polycrystalline samples usually 'have textures', i.e., the distributions of crystallite orientations are not random. Also the orientation relationships between neighboring grains frequently show some preferences due to non–random orientation distribution or because of affinity between certain orientations.

Crystallographic texture of a particular polycrystalline sample is determined by its chemical composition and the history of the sample. For example, in plastic deformation, the list of essential factors may include: the initial (pre–deformation) texture, type of deformation (e.g. tension, compression, torsion, rolling), the so–called deformation path (texture changes are not reversible) and the sample's thermal history (temperatures and times). Perfect single crystals have an extreme kind of crystallographic texture with just one orientation present.

9.1 Texture Components in Cubic–Orthorhombic Case

Most crystal structures exhibit nontrivial symmetries and this is an important factor in the investigation of their textures. However, the analysis of a particular texture is influenced not only by crystal symmetry but also by statistical symmetry of the sample. In many cases, one must deal with the orientation relationship between two symmetric objects: the crystal and the sample. Most frequently encountered sample symmetries are orthorhombic, axial, monoclinic and triclinic (no symmetry). Due to the symmetries, non–random orientation distributions are multimodal in the complete orientation domain.

Some well established standards for communication of results exist for fcc and bcc structures, especially for the case of materials with orthorhombic sample symmetry which includes rolled materials. Conventionally, Cartesian sample coordinate system has e_1 along the rolling direction and e_2 perpendicular to the rolling plane.

Textures of real materials are characterized by specifying the location of local maxima of orientation distributions and the 'spread' of orientations around the maxima. The minimal value of an orientation distribution is sometimes referred to as a 'background' or a 'random texture component'.

Table 9.1 contains orientations most frequently encountered in the analysis of materials with cubic-orthorhombic symmetry. Locations of some of these orientations in the box parameterized by Eular angles are displayed in Fig. 9.1.

Table 9.1. Common orientations for cubic-orthorhombic symmetry. Only one set of Euler angles is given. The complete list of equivalent sets would cotain up to $24 \times 4 = 96$ items.

Component	Miller indices	Euler angles $(\varphi_1, \phi, \varphi_2)$
'cube'	$\{0\,0\,1\} <1\,0\,0>$	$(0°, 0°, 0°)$
'brass'	$\{0\,1\,1\} <2\,\bar{1}\,1>$	$(35.264°, 45°, 0°)$
'copper'	$\{1\,1\,2\} <\bar{1}\,\bar{1}\,1>$	$(90°, 35.264°, 45°)$
'S'	$\{2\,1\,3\} <\bar{3}\,\bar{6}\,4>$	$(58.980°, 36.699°, 63.435°)$
'Goss'	$\{0\,1\,1\} <1\,0\,0>$	$(0°, 45°, 0°)$
'shear'	$\{0\,0\,1\} <1\,\bar{1}\,0>$	$(45°, 0°, 0°)$
'Dillamore' (or 'Taylor')	$\{4\,4\,11\} <\overline{11}\,\overline{11}\,8>$	$(90°, 27.215°, 45°)$
	$\{1\,1\,1\} <\bar{1}\,\bar{1}\,2>$	$(90°, 54.736°, 45°)$
	$\{1\,1\,2\} <1\,\bar{1}\,0>$	$(0°, 35.264°, 45°)$
	$\{1\,1\,3\} <1\,\bar{1}\,0>$	$(0°, 25.239°, 45°)$

Orientations of twins

An important role in analysis of recrystallization and deformation textures is played by twinning. Knowing the orientation relationship between a crystal and its twin, orientations of twins to that crystal can be determined from its orientation.

The most common twins of fcc materials are related to the original matrix orientation via the rotation by 60° about <1 1 1> axis. Such twinning of a grain in a general orientation leads to four new nonequivalent orientations. In special cases,

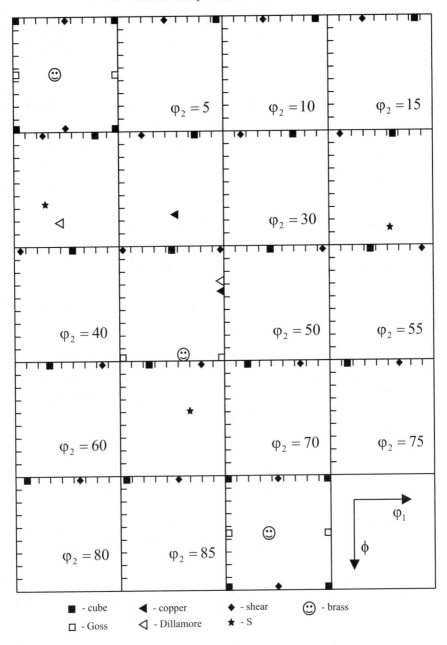

Fig. 9.1. Approximate locations of typical orientations in the "space" of Euler angles for the cubic–orthorhombic case. The ranges of all three angles are from 0° to 90°.

some of these new orientations may overlap or may be identical with the initial orientation. Table 9.2 contains orientations of twins to crystallites in 'cube', 'brass' and 'copper' orientations. In all three cases some overlapping occurs. ☒

Table 9.2. Orientations of twins to 'cube', 'brass' and 'copper' for cubic-orthorhombic symmetry.

Component	Miller indices	Euler angles $(\varphi_1, \phi, \varphi_2)$
Twins to 'cube'	$\{2\,2\,\bar{1}\}$ $<\bar{1}\,2\,2>$	$(135°, 109.471°, 45°)$
(one distinct	$\{\bar{2}\,2\,\bar{1}\}$ $<\bar{1}\,\bar{2}\,2>$	$(225°, 109.471°, 315°)$
orientation)	$\{2\,\bar{2}\,\bar{1}\}$ $<\bar{1}\,\bar{2}\,2>$	$(45°, 109.471°, 135°)$
	$\{\bar{2}\,\bar{2}\,\bar{1}\}$ $<\bar{1}\,2\,\bar{2}>$	$(315°, 109.471°, 225°)$
Twins to 'brass'	$\{4\,1\,1\}$ $<\bar{2}\,7\,1>$	$(171.951°, 76.367°, 75.964°)$
(three distinct	$\{\bar{4}\,1\,1\}$ $<\bar{2}\,1\,\bar{7}>$	$(258.578°, 76.367°, 284.036°)$
orientations)	$\{0\,1\,\bar{1}\}$ $<2\,\bar{5}\,5>$	$(105.793°, 135°, 180°)$
	$\{0\,\bar{1}\,\bar{1}\}$ $<2\,1\,\bar{1}>$	$(324.736°, 135°, 180°)$
Twins to 'copper'	$\{5\,5\,2\}$ $<1\,1\,\bar{5}>$	$(270°, 74.207°, 45°)$
(three distinct	$\{\bar{7}\,1\,\bar{2}\}$ $<1\,5\,\bar{1}>$	$(348.463°, 105.793°, 278.130°)$
orientations)	$\{1\,\bar{7}\,\bar{2}\}$ $<5\,1\,\bar{1}>$	$(191.537°, 105.793°, 171.870°)$
	$\{\bar{1}\,\bar{1}\,\bar{2}\}$ $<\bar{1}\,\bar{1}\,1>$	$(90°, 144.736°, 225°)$

Besides texture 'components' with a spread of crystallites' orientations around the center of the component, the case with orientations scattered along a line in the orientation space is frequently encountered . These structures are referred to as texture fibers (see, e.g., Hirsch and Lücke, 1988). The most common fibers are listed in Table 9.3.

9.2 Rational Orientation Relationships

Misorientations are usually analyzed in connection with observations of phase transformations, precipitation of particles, growth of lamellar eutectics, epitaxial growth, deposition of solid thin films on (poly)crystalline substrates et cetera. Traditionally, interphase misorientations are referred to as orientation relationships. Considerations based on thermodynamics or molecular dynamics are usually applicable for explaining the presence of particular relationships. In more crude explanations, it is assumed that close–packed arrays of atoms tend to be parallel, or close–packed planes of two different structures tend to be parallel, or there is an affinity between cross–sections exposing planes with similar interplanar spacing.

Orientation relationships between structures with different phases are customarily specified by Miller indices of parallel planes or/and directions. Frequently, the notation not only carries information about the misorientation, but also indicates crystallography (inclination) of the actual interface between two phases. For non–enantiomorphic crystals, a pair of planes and a pair of directions are sufficient for unambiguous determination of a relationship if the directions are not perpendicular to the planes. Besides that, the choice

Table 9.3. Texture fibers for cubic-orthorhombic symmetry; (i) characterization of the fiber, (ii) location in the standard domain shown in Fig. 9.1, (iii) characteristic components located on the fiber.

Fiber	Specification
α	(i) <110> parallel to the normal direction
	(ii) $(\varphi_1, \phi, \varphi_2) = (\varphi_1, 45°, 0°)$ or $(\varphi_1, 45°, 90°)$ or $(\varphi_1, 90°, 45°)$
	(iii) 'Goss', 'brass'
β	(i) from 'brass' through 'S' to 'copper' texture component
	(ii) there is no unique axis of rotation
	(iii) 'brass', 'S', 'copper'
γ	(i) <111> parallel to the normal direction
	(ii) $(\varphi_1, 54.736°, 45°)$
	(iii) $\{1\,1\,1\} <\bar{1}\,\bar{1}\,2>$
η	(i) <100> parallel to the rolling direction
	(ii) $(90°, 90°, \varphi_2)$ or $(0°, \phi, 90°)$
	(iii) 'cube', 'Goss'
τ (or ε)	(i) <110> parallel to the transverse direction
	(ii) $(90°, \phi, 45°)$ or symmetrically equivalent line, which can be expressed as: $\varphi_1 = \arccos(1/\sqrt{1 + \cos^2 t})$, $\phi = \arccos(\sin t/\sqrt{2})$ and $\varphi_2 = \arcsin(\sin t/\sqrt{1 + \cos^2 t})$, where $t \in [0, \pi/2]$.
	(iii) 'copper', 'Dillamore', twin to 'copper', 'Goss', 'shear', $\{1\,1\,1\} <\bar{1}\,\bar{1}\,2>$

of planes or directions is arbitrary, and two relations which at first appear as different, may actually correspond to one orientation relationship. A few of the most commonly encountered relationships are listed below..

Relationships between fcc and bcc structures
Probably the best known orientation relationships are those referred to as Kurdjumov–Sachs and Nishiyama–Wassermann relationships. The Kurdjumov–Sachs orientation relationship can be written as

$$(111)_{fcc} \parallel (\bar{1}10)_{bcc}$$
$$[1\bar{1}0]_{fcc} \parallel [111]_{bcc} \quad .$$

Brief calculation shows that the coordinate systems of the two structures are brought into coincidence by the 90° rotation about the common $[11\bar{2}]$ axis. Nishiyama–Wassermann orientation relationship has the form

$$(111)_{fcc} \parallel (011)_{bcc}$$
$$[01\bar{1}]_{fcc} \parallel [100]_{bcc} \quad .$$

This misorientation differs from the relationship of Kurdjumov–Sachs by 5.26°. Sometimes, the second part of the Nishiyama–Wassermann orientation relationship is equivalently expressed as $[\bar{2}11]_{fcc} \parallel [01\bar{1}]_{bcc}$.

The orientation relationship of Pitsch (1959)

$$(110)_{fcc} \parallel (\bar{1}\bar{1}2)_{bcc}$$
$$[001]_{fcc} \parallel [1\bar{1}0]_{bcc}$$
$$[1\bar{1}0]_{fcc} \parallel [\bar{1}\bar{1}1]_{bcc} \quad .$$

can be written in the form $(011)_{fcc} \parallel (111)_{bcc}$ and $[100]_{fcc} \parallel [01\bar{1}]_{bcc}$. This explains why it is also referred to as the "inverse Nishiyama–Wassermann" orientation relationship.

Bain (or Baker–Natting) orientation relationship between fcc and bcc structures is relatively simple

$$(001)_{fcc} \parallel (001)_{bcc}$$
$$[110]_{fcc} \parallel [100]_{bcc} \quad .$$

The coordinate systems of the two structures are brought into coincidence by the $45°$ rotation about the common $[001]$ axis. ⊠

Relationships between bcc and hcp structures
Orientation relationships between bcc and hcp structures are also frequently considered. Probably the best known among them is Burgers orientation relationship observed in *bcc* to *hcp* phase transformations in zirconium, titanium and their alloys

$$(0001)_{hcp} \parallel (101)_{bcc}$$
$$[110]_{hcp} \parallel [11\bar{1}]_{bcc} \quad .$$

The Miller direction indices (not Weber indices) for hcp structure are being used.
The orientation relationship of Pitsch–Schrader is usually given as

$$[110]_{hcp} \parallel [100]_{bcc}$$
$$[\bar{1}11]_{hcp} \parallel [010]_{bcc}$$
$$[1\bar{1}1]_{hcp} \parallel [001]_{bcc} \quad .$$

However, the angles between directions on the hcp side of the Pitsch–Schrader relation are fully compatible with the angles between directions on the bcc side (which obviously are $90°$) only if $c/a = \sqrt{3}$. In other cases, with fixed $[110]_{hcp} \parallel [100]_{bcc}$ and the small deviations between two remaining pairs equal, the relationship can be written as

$$(0001)_{hcp} \parallel (011)_{bcc}$$
$$[110]_{hcp} \parallel [100]_{bcc} \quad ;$$

this form was given by Dyson *et al.* (1966). ⊠

One additional case
There are a large number of various orientation relations considered in literature and applicable to some specific materials. We would like to mention the frequently considered case of relationships in steels between cementite (which is orthorhombic) and ferrite (bcc) or austenite (fcc):
Bagaryatsky relationship

$$[100]_{cem} \parallel [0\bar{1}1]_{bcc}$$
$$[010]_{cem} \parallel [1\bar{1}\bar{1}]_{bcc}$$
$$[001]_{cem} \parallel [211]_{bcc} \quad ,$$

Isaichev relationship

$$(\bar{1}03)_{cem} \parallel (011)_{bcc}$$
$$[010]_{cem} \parallel [\bar{1}1\bar{1}]_{bcc} \quad ,$$

Pitsch relationship

$$(100)_{cem} \parallel (5\bar{5}4)_{fcc}$$
$$(010)_{cem} \parallel (110)_{fcc}$$
$$[001]_{cem} \parallel [\bar{2}25]_{fcc} \quad .$$

See Bhadeshia (2001) and references given there for more on this subject. ⊠

Impractical but more universal way of specifying orientation relationships between two lattices was proposed by Fortes (1984). Let a_i' and a_i'' be bases of the two lattices. The relationship between them can be written as

$$a_i'' = a_j' Q^j{}_i \; .$$

Even with the standardized choice of the bases, the matrix Q of a given orientation relationship is not unique because of lattice symmetry. But it is 'less arbitrary' than the specification by parallel planes and directions. Calculation of Q from the traditional expression of the relationship is a bit cumbersome but straightforward.

Let two relations of parallelism between some directions and/or planes be given. Parallelism between directions $[u'^1 \, u'^2 \, u'^3] \parallel [u''^1 \, u''^2 \, u''^3]$ can be expressed as $u'^i a_i' \propto u''^i a_i''$ and parallelism between planes $[h_1' \, h_2' \, h_3'] \parallel [h_1'' \, h_2'' \, h_3'']$ corresponds to $h'^i a_i' \propto h''^i a_i''$, where $h'^i = g'^{ij} h_j'$, $h''^i = g''^{ij} h_j''$, and g'^{ij}, g''^{ij} are contravariant metric matrices of the two lattices (i.e., inverses to the metric matrices $\{g_{ij}'\} = a_i' \cdot a_j'$ and $\{g_{ij}''\} = a_i'' \cdot a_j''$, respectively; see sections 2.8 and 2.9). Hence, one has two relations of the form $v'^i a_i' \propto v''^i a_i''$, where v'^i and v''^i represent indices of either directions or planes. A third relation of that type can be added by calculating vector products $(u'^i a_i') \times (h'^i a_i')$ and $(u''^i a_i'') \times (h''^i a_i'')$. [1] The vectors $v'^i a_i'$ and $v''^i a_i''$ are then normalized using $v'^i / \sqrt{g_{kl}' v'^k v'^l} \to v'^i$ and $v''^i / \sqrt{g_{kl}'' v''^k v''^l} \to v''^i$, and the proportionality can be replaced by equality. With an additional index k $(= 1, 2, 3)$ enumerating the equations,

$$v'^i{}_k a_i' = v''^i{}_k a_i'' \; .$$

Since $a_i' \cdot a_j'' = Q^i{}_j$, the above relation can be written as $v'^i{}_k Q^i{}_j = v''^i{}_k g_{ij}''$ (or $v'^i{}_k g_{ij}' = v''^i{}_k Q^j{}_i$) and the matrix Q can be obtained by taking the inverse of v' (or v''); in brief notation

$$Q = (g'' v'' v'^{-1})^T = g' v' v''^{-1} \; . \tag{9.1}$$

If $g' = g''$, the matrix Q satisfies (2.57) and represents a rotation.

⊡ *Example*
With $a_i^{bcc} = a_j^{fcc} Q^j{}_i$, the matrices Q for the Kurdjumov–Sachs (KS) and Nishiyama-Wassermann (NW) relationships are

$$Q_{KS} = \frac{1}{6} \begin{bmatrix} 1 & 1 + 2\sqrt{6} & -2 + \sqrt{6} \\ 1 - 2\sqrt{6} & 1 & -2 - \sqrt{6} \\ -2 - \sqrt{6} & -2 + \sqrt{6} & 4 \end{bmatrix}, \quad Q_{NW} = \frac{1}{6} \begin{bmatrix} 0 & -2\sqrt{3} + \sqrt{6} & 2\sqrt{3} + \sqrt{6} \\ 3\sqrt{2} & \sqrt{3} + \sqrt{6} & -\sqrt{3} + \sqrt{6} \\ -3\sqrt{2} & \sqrt{3} + \sqrt{6} & -\sqrt{3} + \sqrt{6} \end{bmatrix} .$$

These matrices were calculated directly from Miller indices given above and they do not represent rotations with the smallest angles. In order to get the smallest rotation angles, the matrices are transformed (using symmetry operations (6.3)) to

$$Q_{KS} = \frac{1}{6} \begin{bmatrix} 1 + 2\sqrt{6} & 2 - \sqrt{6} & 1 \\ 1 & 2 + \sqrt{6} & 1 - 2\sqrt{6} \\ 2 - \sqrt{6} & 4 & 2 + \sqrt{6} \end{bmatrix}, \quad Q_{NW} = \frac{1}{6} \begin{bmatrix} 2\sqrt{3} + \sqrt{6} & 0 & 2\sqrt{3} - \sqrt{6} \\ -\sqrt{3} + \sqrt{6} & 3\sqrt{2} & -\sqrt{3} - \sqrt{6} \\ \sqrt{3} - \sqrt{6} & 3\sqrt{2} & \sqrt{3} + \sqrt{6} \end{bmatrix} .$$

---

[1] The components of the vector product of in non–Cartesian crystallographic system with the metric g' are given by $\sqrt{|\det(g')|} \, g'^{ij} \varepsilon_{jkl} u'^k h'^l$.

The axes and the smallest rotation angles obtained from these matrices are

$$((0.9679, 0.1776, 0.1776), 42.848°) \text{ and } ((0.9761, 0.2007, 0.0831), 45.988°)$$

for the Kurdjumov–Sachs and Nishiyama-Wassermann orientation relationships, respectively. See also Fig. 9.3. $\boxtimes$

The above construction will help us to show that attention to details is required when dealing with relationships expressed through parallel planes and directions. Let us take a simple but instructive example with two cubic structures denoted by 1 and 2. Let the relationship between them be described as

$$(111)_1 \parallel (111)_2$$
$$[1\bar{1}0]_1 \parallel [\bar{1}10]_2 \ .$$

From (9.1),

$$Q = \frac{1}{3} \begin{bmatrix} -1 & 2 & 2 \\ 2 & -1 & 2 \\ 2 & 2 & -1 \end{bmatrix} ,$$

which is the matrix representing the rotation by 180 degrees about [111] direction of any of the two structures. If both of them were fcc, that would be the misorientation between twins. Now, if the direction $[\bar{1}10]_2$ in the second line defining the relationship is replaced by the opposite but eqivalent direction $[1\bar{1}0]_2$ we get the trivial relationship

$$(111)_1 \parallel (111)_2$$
$$[1\bar{1}0]_1 \parallel [1\bar{1}0]_2 \ ,$$

which would correspond to no difference in orientation for the same fcc on both sides of the relations.

9.3 Coincident Sublattices

The concept of coincident sublattices (CSLs)[2] frequently appears in connection with models of crystalline interfaces, especially models of grain boundaries. For some particular misorientations, the lattices of neighboring crystallites contain a common sublattice which may lead to a periodic atomic structure of the interface, and in turn to special physical properties. Beyond the highest level of coincidence (low angle boundaries and experimentally observable twins), the subject is somewhat controversial. However, CSL misorientations get so much attention that we decided to include a note on this subject. The book by Randle (1996) is an overview of CSL boundaries in cubic materials from the perspective of materials scientist.

Let a_i $(i = 1, 2, 3)$ be linearly independent vectors. As was already mentioned, the lattice Λ is a set of linear combinations of the form $k_i a_i$ with integer k_i. A sublattice of Λ is a lattice which, as a set, is a subset of Λ. Let M be a non–singular matrix with integer entries; the lattice with the basis

---

[2] The abbreviation CSL stands for the unfortunate name Coincident Site Lattice; the spurious term 'site' is at odds with the (atomic) 'site' used in crystallography.

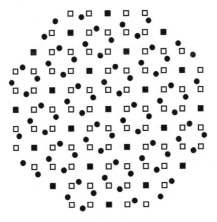

Fig. 9.2. Schematic illustration of coincident lattices. Every fifth disc overlaps a square.

$b_i = M^j{}_i a_j$ is a sublattice of Λ. Opposite is also true: the matrix relating the bases of a lattice to a basis of its sublattice has integer entries. The fraction of lattice vectors which also belong to the sublattice is $1/\Sigma$, where

$$\Sigma = |\det(M)| \ .$$

If $|\det(M)| = 1$ the sublattice is identical to the lattice.

> **Warning**
> For some lattices, commonly used basis vectors correspond to 'conventional' non–primitive cells. It is assumed here that a_i span a primitive cell. ⊠

Let Λ' and Λ'' be two lattices with bases a_i' and a_i'', respectively. Λ' and Λ'' have sublattices which are coincident only if there exist matrices N' and N'' having integer entries and satisfying $N'^j{}_i a_j' = N''^j{}_i a_j''$. This means that lattice bases are related by $a_i'' = Q^j{}_i a_j'$, where the matrix $Q := N''^{-1} N'$ has rational entries. A rotation may be applied to one of the lattices to bring the sublattices into coincidence; if the basis a_i'' is displaced by the rotation R,

$$R^j{}_i a_j'' = Q^j{}_i a_j' \ .$$

Due to Eq.(2.57) ($g_{ij}'' = R^k{}_i R^l{}_j g_{kl}''$), the metrics of the lattices are related through

$$g_{ij}'' = g_{kl}' Q^k{}_i Q^l{}_j \ . \tag{9.2}$$

This expression is invariant of R.

The main problem of the theoretical investigation of CSLs is to solve equation (9.2) with respect to Q having rational entries. The next task is to decompose it into the product $Q = N''^{-1} N'$ with N' and N'' having integer entries, i.e., to determine the actual coinciding sublattices. We want to know N' and N'' with the smallest possible $\Sigma' = |\det(N')|$ and $\Sigma'' = |\det(N'')|$ and corresponding to the highest level of coincidence. Let k' be the smallest positive integer such that $k'Q^{-1}$ has integer entries, and let k'' be the smallest positive integer such that $k''Q$ has integer entries. It can be proved (Fortes, 1983) that for three–dimensional lattices, the highest level of coincidence occurs for Σ' given as the smallest positive integer such that both

$$\Sigma'/k' \quad \text{and} \quad \Sigma'/(k'' \, |\det(Q)|)$$

are integers. Since $\Sigma'' \, |\det(Q)| = \Sigma'$, the second constraint is equivalent to the condition that Σ''/k' is an integer.

The property of sublattice coincidence is transitive: if Λ' has a sublattice coincident with a sublattice of Λ'', and Λ'' has a sublattice coincident with a sublattice of Λ''', then Λ' has a sublattice coincident with a sublattice of Λ'''. This follows directly from the observation that the product of non–singular matrices having rational entries is a non–singular matrix with rational entries. The above property is manifested at interface triple junctions. If two interfaces separate lattices having coincident sublattices, also the third interface separates lattices with coincident sublattices.

DSC lattice

If Λ is a sublattice of another lattice, the latter is called a superlattice of Λ. Let M be a non–singular matrix with integer entries; the lattice with the basis $b_i = (M^{-1})^j{}_i a_j$ is a superlattice of Λ.

Let Λ' and Λ'' be two lattices with bases a_i' and a_i'', respectively. There is a superlattice common to both Λ' and Λ'' only if there exist matrices N' and N'' having integer entries and satisfying $(N'^{-1})^j{}_i a_j' = (N''^{-1})^j{}_i a_j''$. This means that lattice bases are related by $a_i'' = Q^j{}_i a_j'$, where the matrix $Q = N''N'^{-1}$ has rational entries. Analogously to coincident sublattices, the metrics of the lattices having a common superlattice must be related through Eq.(9.2). For given two lattices, their 'smallest' common superlattice is called DSC lattice. ⊠

Relationship between DSC and CSL

Let a_i $(i = 1, 2, 3)$ be the basis of Λ. The lattice Λ^* reciprocal to Λ is based on vectors a^{*i} satisfying $a_i \cdot a^{*j} = \delta_i^j$. It is relatively easy to show that the lattice reciprocal to CSL of Λ' and Λ'' is the DSC lattice of Λ'^* and Λ''^*. ⊠

If $g'' = g' =: g$, the matrix Q satisfies Eq.(2.57), i.e., it is a rotation matrix in the system with metric g. Since $|\det(Q)| = 1$, one has $\Sigma'' = \Sigma' =: \Sigma$, where Σ is the smallest positive integer such that both ΣQ and ΣQ^{-1} have integer entries. Only that case will be considered further.

9.3.1 Primitive cubic lattice

For the cubic primitive lattice, the metric can be chosen as $g_{ij} = a\delta_{ij}$, where a is the lattice parameter. The parameter a is a factor on both sides of 9.2 and is reduced. Thus, the equation does not involve any physical constants. The sought Q is simply an orthogonal matrix with rational entries. Since every lattice is centrosymmetric, the search for Q can be limited to *special* orthogonal matrices. In the considered case of the cubic primitive lattice, the parameter Σ takes only odd values.

Proof
To see that, let $M^i{}_j := \Sigma Q^i{}_j$. Hence, $M^k{}_i M^k{}_j = \Sigma^2 \delta_{ij}$. The explicit form of the entry for $i = w = j$ is $M^k{}_{(w)} M^k{}_{(w)} = \Sigma^2$. Let us assume that Σ is even: $\Sigma = 2\sigma$ and σ is an integer. $M^k{}_{(w)}$ can be expressed as $M^k{}_{(w)} = 2L_k + \lambda_k$, where L_k are integers, and each of λ_k equals either 0 or 1. The relation $M^k{}_{(w)} M^k{}_{(w)} = \Sigma^2$ can be written as $\lambda_k \lambda_k = 4(\sigma^2 - L_k L_k - L_k \lambda_k)$. Hence, all λ_k must be 0 and it must occur for all w; but this would mean that Σ and all entries of M would be even, and Σ would not be the smallest positive integer such that ΣQ has integer entries. Thus, Σ cannot be even. $\boxtimes$

In order to determine CSL orientation relationships, it is convenient to use the parameterization by r^i described in section 2.2. Based on the relation (2.13) $(r^i = -\varepsilon_{ijk} Q^j{}_k/(1 + Q^l{}_l))$, if $Q^k{}_l$ are rational, so are the parameters r^i. Let l^0 be the smallest positive integer such that all $l^i = l^0 r^i$ are integers. Because of (2.14),

$$l^\mu l^\mu Q^i{}_j = ((l^0 l^0 - l^k l^k)\delta_{ij} + 2l^i l^j - 2\varepsilon_{ijk} l^k l^0) , \quad \mu = 0, ..., 3 . \tag{9.3}$$

Thus, all entries of the matrix $l^\mu l^\mu Q$ are integers, and $l^\mu l^\mu$ is a multiple of Σ, i.e., $m\Sigma = l^\mu l^\mu$, where m is a positive integer. Hence, CSL orientation relationships can be deduced from the decompositions of $m\Sigma$ into $l^\mu l^\mu \neq 0$. With a given set of l^μ such that $m\Sigma = l^\mu l^\mu$, one can calculate $r^i = l^i/l^0$, then the rotation matrix (from (2.14)) and Σ. To aviod the inconvenience of the case $l^0 = 0$, the matrix $m\Sigma Q$ can be determined directly from (9.3). The factor m is the largest common divisor of $l^\mu l^\mu$ and all of $(l^0 l^0 - l^k l^k)\delta_{ij} + 2l^i l^j - 2\varepsilon_{ijk} l^k l^0$.

The CSL misorientations are usually specified by giving the rotation axis and the rotation angle. The axis is given by $[l^1, l^2, l^3]$, and the angle can be calculated from $\tan(\omega/2) = \sqrt{r^i r^i}$ or from Q by using Eqs. (2.18) and (2.19). Many misorientations obtained that way will be equivalent due to lattice symmetries. Usually, a misorientation with the smallest rotation angle is used to represent the set of equivalent misorientations.

As an example, let us take the numbers $l^0 = 2$, $l^1 = 1$, $l^2 = 0 = l^3$. They lead to $r^1 = 1/2$ and $r^2 = 0 = r^3$, and subsequently, by the use of (2.14), to

$$Q = \frac{1}{5} \begin{bmatrix} 5 & 0 & 0 \\ 0 & 3 & -4 \\ 0 & 4 & 3 \end{bmatrix} .$$

This gives $\Sigma = 5$ for the rotation about $[l^1, l^2, l^3] = [100]$ by $2\arctan(\sqrt{r^i r^i}) = 2\arctan(1/2)$ ($\sim 53.13°$). However, this is not the smallest rotation angle leading to $\Sigma = 5$ coincidence. The values $l^0 = 3$, $l^1 = 1$, $l^2 = 0 = l^3$ give an equivalent misorientation with the axis $[100]$ and the angle equal to $36.87°$. It is listed in Table 9.4 along with other low Σ CSL misorientations. (See also Fig. 9.3.) If there are numerous non–equivalent misorientations corresponding to the same value of Σ, letters of the Latin alphabet are used to discriminate between them.

It can be shown that the CSL misorientations for face centered cubic and body centered cubic lattices, with axes expressed in the conventional non–primitive cells, are identical with those for the primitive cubic lattice.

Relationships for DSC lattices
Since the lattice reciprocal to the primitive cubic lattice is also primitive cubic, the DSC orientation relationships of the primitive cubic lattice are the same as CSL relationships. Moreover, because the lattice reciprocal to the face centered cubic lattice is the body centered cubic lattice and *vice versa*, then their DSC orientation relationships are the same as their CSL relationships. ⊠

Table 9.4. Low Σ CSL orientation relationships for cubic primitive lattice.

Σ	Angle	Axis
1	0.0000°	
3	60.0000°	<111>
5	36.8699°	<100>
7	38.2132°	<111>
9	38.9424°	<110>
11	50.4788°	<110>
13a	22.6199°	<100>
13b	27.7958°	<111>
15	48.1897°	<210>
17a	28.0725°	<100>
17b	61.9275°	<221>
19a	26.5254°	<110>
19b	46.8264°	<111>
21a	21.7868°	<111>
21b	44.4153°	<211>
23	40.4591°	<311>
25a	16.2602°	<100>
25b	51.6839°	<331>
27a	31.5863°	<110>
27b	35.4309°	<210>
29a	43.6028°	<100>
29b	46.3972°	<221>
31a	17.8966°	<111>
31b	52.2003°	<211>
33a	20.0500°	<110>
33b	33.5573°	<311>
33c	58.9924°	<110>

Brandon criterion
Coherent boundaries, by definition, are in exact CSL relationships. For other crystallite pairs such relationship may be satisfied only approximately. In those cases, the criterion of Brandon is in use: a crystallite misorientation is classified as a CSL type if the angular deviation between the misorientation and the exact CSL relationship does not exceed certain Σ–dependent limit, usually $15°/\sqrt{\Sigma}$. ⊠

9.3.2 The general case

It is easy to notice that the rationality of numbers is critical for the results concerning CSLs. On the other hand, non–discrete physical constants, like

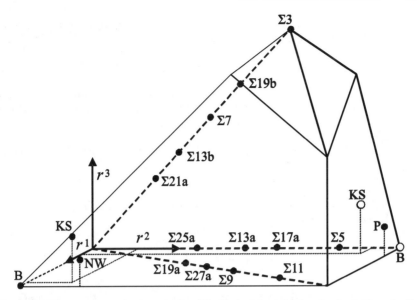

Fig. 9.3. The asymmetric domain for (O,O) (cubic–cubic) misorientations in Rodrigues space with locations of some CSL misorientations. A rotation representing a CSL is equivalent to its inverse, half of the domain for (O,O) is sufficient for showing CSL relationships. Such reduction is not allowed if orientation relationships between different phases are considered. The locations of Kurdjumov–Sachs (KS), Nishiyama-Wassermann (NW), Bain (B) and Pitsch (P) orientation relationships are also shown.

cell parameters, cannot be classified as rational or irrational. For non–cubic systems, the equation $g_{ij} = Q^k{}_i Q^l{}_j g_{kl}$ does contain cell parameters and the CSL theory of grain boundaries somewhat looses its appeal.

In the general case, the CSL orientation relationships can be determined in a way similar to that applicable to the cubic primitive lattice. However, relations appropriate for non–Cartesian coordinate systems must be used. Thus, instead of (2.13) and (2.14) we need (2.60) and (2.61), respectively.

If the metric g has rational entries, the approach is analogous to the case of the cubic primitive lattice. Based on the relation (2.60), if $Q^m{}_l$ and g_{tm} are rational, so are the parameters $\varrho^s = -\varepsilon^{stl} g_{tm} Q^m{}_l / (1 + Q^u{}_u)$. Let σ be the smallest positive integer such that the entries of σg_{ij} are integers, and let l^0 be the smallest positive integer such that all $l^i = l^0 \varrho^i$ are integers. Because of (2.61),

$$\sigma^3 (\det(g) l^0 l^0 + g_{st} l^s l^t) Q^k{}_l$$
$$= \sigma^3 ((\det(g) l^0 l^0 - g_{uw} l^u l^w) \delta^k{}_l + 2 g_{tl} l^k l^t - 2 \det(g) \varepsilon_{lnm} g^{mk} l^0 l^n) \tag{9.4}$$

with the right side taking only integer values. Thus, also $\sigma^3 (\det(g) l^0 l^0 + g_{st} l^s l^t) Q^k{}_l$ are integers. The matrix Q^{-1} inverse to Q satisfies $(Q^{-1})^i{}_j = g_{jk} Q^k{}_l g^{li}$. Hence, its relation to ϱ differs from (9.4) by the sign of the last term. Therefore, also $\sigma^3 (\det(g) l^0 l^0 + g_{st} l^s l^t) Q^{-1}$ has integer entries. Thus,

$\sigma^3(\det(g)l^0l^0 + g_{st}l^sl^t)$ is a multiple of Σ, and CSL orientation relationships can be deduced from the decompositions

$$m\Sigma = \sigma^3(\det(g)l^0l^0 + g_{st}l^sl^t) \neq 0 ,$$

where m is a positive integer. Starting with four integers l^μ, one can calculate $\varrho^i = l^i/l^0$, the rotation matrix (from Eq.(2.61)) and Σ. To aviod the inconvenience of the case $l^0 = 0$, the matrix $m\Sigma Q$ can be determined directly from (9.4). The factor m is the largest common divisor of $\sigma^3(\det(g)l^0l^0 + g_{st}l^sl^t)$, all elements of the right hand side of (9.4) and of its equivalent corresponding to Q^{-1}.

□ *Example*
Let us consider the hexagonal lattice with the ratio $c/a = \sqrt{8/3}$. Based on (6.1), the metric has the form

$$g = \begin{bmatrix} 1 & -1/2 & 0 \\ -1/2 & 1 & 0 \\ 0 & 0 & 8/3 \end{bmatrix} .$$

Moreover, $\det(g) = 3$ and $\sigma = 6$. As an example, we will take the numbers $l^0 = 3$, $l^1 = 4$, $l^2 = 2$, $l^3 = 0$. Substitution to (9.4) leads to

$$6480Q = \begin{bmatrix} 6480 & 0 & 0 \\ 2592 & 1296 & -10368 \\ -1944 & 3888 & 1296 \end{bmatrix} , \quad 6480Q^{-1} = \begin{bmatrix} 6480 & 0 & 0 \\ 2592 & 1296 & 10368 \\ 1944 & -3888 & 1296 \end{bmatrix} ,$$

This means that

$$Q = \frac{1}{10}\begin{bmatrix} 10 & 0 & 0 \\ 4 & 2 & -16 \\ -3 & 6 & 2 \end{bmatrix} , \quad Q^{-1} = \frac{1}{10}\begin{bmatrix} 10 & 0 & 0 \\ 4 & 2 & 16 \\ 3 & -6 & 2 \end{bmatrix} .$$

Hence, the frequency of coincidence is $1/\Sigma = 1/10$. Moreover, $m = 648$ and $\varrho = [4\,2\,0]/3$. For other low Σ misorientations see Table 9.5 and Warrington (1975). ⊠

If the metric tensor has irrational entries, the choice of misorientation axes may be limited. In the extreme case, the set of possible coincident sublattices may be restricted to $\Sigma = 1$. It is easy to give an example. For coincidence to occur, the quantity $\varepsilon_{ijk}g^{im}\varrho^k$ must be rational since it is given by $\varepsilon_{ijk}g^{im}\varrho^k = -(Q^m{}_j - (Q^{-1})^m{}_j)/(1 + Q^l{}_l)$ and Q has rational entries. For the hexagonal lattice with a normalised to 1 and irrational $(c/a)^2 = c^2$, this condition eliminates (from the list of possible CSLs) all cases with non–zero ϱ^1 or ϱ^2, i.e., only rotations about the c direction lead to coincidence between sublattices.

Table 9.5. Low Σ CSL orientation relationships for hexagonal lattice with $(c/a)^2 = 8/3$.

Σ	Angle	Miller indices of axis
7	21.7868°	<0 0 1>
10	78.4630°	<2 1 0>
11	62.9643°	<2 1 0>
13	27.7958°	<0 0 1>
14	44.4153°	<2 1 0>
17	86.6277°	<1 0 0>
18	70.5288°	<1 0 0>
19	13.1736°	<0 0 1>
22	50.4788°	<1 0 0>
25	23.0739°	<2 1 0>
26	87.7958°	<8 0 1>
27	38.9424°	<1 0 0>
29	66.6372°	<16 0 3>
31a	17.8966°	<0 0 1>
31b	56.7436°	<5 1 0>
34	53.9681°	<4 0 1>
35a	34.0477°	<2 1 0>
35b	57.1217°	<2 1 0>
37a	9.4300°	<0 0 1>
37b	72.7047°	<7 2 0>
38a	26.5254°	<1 0 0>
38b	73.1736°	<16 8 3>
41	55.8767°	<1 0 0>
43a	15.1782°	<0 0 1>
43b	83.3226°	<8 1 0>
45	86.1774°	<3 1 0>

10

Diffraction Geometry

DETERMINATION of crystalline orientations is one of the important issues in investigation of polycrystalline materials. Numerous experimental techniques are used. Those based on diffraction are considered to be the most efficient, accurate and versatile, i.e., applicable to a wide range of materials: form large crystals to nano–scale crystallites. The standard methods include x–ray diffraction, electron diffraction and neutron diffraction. Getting orientations from diffraction data usually requires considerable amount of computations. Our main goal is to give an idea how such computations are performed. Although, we are touching the experimental side of the texture analysis, the reader should consult other sources for descriptions of actual measurements.

The most essential application of diffraction in crystal research is crystal structure determination. This however, is not a subject of texture analysis. The problem here is considerably simpler because the crystal structure is assumed to be known. Orientations of individual crystallites are usually determined based on the geometry of diffraction without considering radiation intensities. For multiphase materials, this can be extended to phase discrimination; if a number of phases are present, the additional task is to indicate the phase to which a given pattern corresponds.

The diffraction techniques based on a narrow beam (comparing to the grain size) allow determination of orientations of individual crystallites. With such local orientation measurements, orientation maps can be created (e.g., Inokuti, Maeda & Ito, 1985). Orientation mapping techniques require large numbers of orientations and extensive computations. It must be also noted that the task of determining orientations of individual crystallites is challenging in the case of small grain size or large residual strain, when it is difficult to get reasonably good diffraction patterns.

If only orientation distribution is required, it can be calculated with a very good approximation from experimentally obtained pole figures. Pole figures can be measured for single crystals. However, more common is the measurement with each point of a pole figure containing information about orientations of numerous crystallites. The problem of getting orientation distributions form pole figure data of that type is also computationally non–trivial.

10.1 Elementary Relations

Diffraction of x-rays and electrons is strictly an interference phenomenon but can be seen as a reflection from crystallographic planes. The geometry of diffraction is governed by the Laue equation

$$\pm\mathbf{g} = \mathbf{k} - \mathbf{k}_0 \text{ , with } \mathbf{k}\cdot\mathbf{k} = 1/\lambda^2 = \mathbf{k}_0\cdot\mathbf{k}_0 \text{ ,} \qquad (10.1)$$

where λ is the radiation wavelength, the wave vectors $\mathbf{k}_0$ and $\mathbf{k}$ indicate directions of the incident and reflected beams, respectively, and the scattering vector $\mathbf{g}$ is a vector of the crystal reciprocal lattice (and is normal to the diffracting plane).[1] The fact that the sign of $\mathbf{g}$ cannot be determined from $\mathbf{k}_0$ and $\mathbf{k}$ is referred to as Friedel's law. Because of the discrete nature of the reciprocal lattice, a diffraction pattern is obtained only if some quantities involved are variable. For instance, in the Laue method, by the use of polychromatic radiation, λ covers certain range of spectrum. In techniques with convergent incident rays (channeling patterns, electron backscattered patterns, Kikuchi patterns, convergent beam electron diffraction, x-ray Kossel patterns) the direction $\mathbf{k}_0$ of the incident beam covers certain solid angle. In conventional pole figure measurements the variation of the direction of $\mathbf{k}_0$ is achieved by rotating the sample.

Intensity distribution around points of reciprocal lattice
Although the Laue equation is based on discrete reciprocal lattice, there is always a spread of radiation intensities around points of the lattice. For example, for spot patterns in transmission electron microscopy, that spread is large because of the thinness of the sample. It makes a spot visible for a range of sample orientations. The effect is good for perception and analysis of patterns but on the other hand, it causes low precision of orientation determination. ⊠

With known crystal structure, the scattering vectors of feasible reflections can be relatively easily calculated via structure factor. Let such calculated scattering vector be denoted by $\mathbf{h}$. The relation of its Cartesian coordinates (h_1, h_2, h_3) to Miller indices $(h\ k\ l) = (h_1^* \ h_2^* \ h_3^*)$ of the reflecting plane is given by

$$h_i^* = A^j{}_i h_j \text{ ,}$$

where $A^j{}_i$ is the j-th coordinate of the i-th basis vectors of the direct lattice in the Cartesian coordinate system; see Eq.(2.63). This formula is valid for non–conventional unit cells as long as the Miller indices of reflecting planes are consistent with the choice of a cell.

10.2 Orientations of Individual Crystallites

For determining orientations of individual crystals, geometric features of the diffraction pattern in the sample reference frame are matched with the same features obtained from crystallographic data. In particular, reciprocal lattice

---

[1] Since the last two chapters have more 'physical' character, from now on we will follow the convention in which vectors and tensors are denoted by boldface letters.

vectors ($\mathbf{g}$) acquired from the experimental diffraction pattern are fit to vectors ($\mathbf{h}$) calculated from crystallographic parameters.

The first task is to determine the normal $\mathbf{g}$ from the location of the pattern features registered on a detector. The procedure depends on the type of experimental method and particular experimental setup. Of all fore–mentioned experimental methods, those with convergent incident rays are currently the most frequently used and will serve here as an example. In this case, the direction of the vector $\mathbf{k}_0$ is arbitrary. The possible directions of $\mathbf{k}$ can be obtained from (10.1). After taking the scalar product of its sides, one has $g^2 = \mathbf{g} \cdot \mathbf{g} = 2(1/\lambda^2 - \mathbf{k} \cdot \mathbf{k}_0)$. On the other hand, by taking the scalar product of the Laue equation and $\mathbf{k}$, we obtain $\pm \mathbf{k} \cdot \mathbf{g} = 1/\lambda^2 - \mathbf{k} \cdot \mathbf{k}_0$. These two relations lead to the following condition on $\mathbf{k}$ (Ryder & Pitsch, 1968)

$$\pm \mathbf{k} \cdot \mathbf{g} = g^2/2 \ . \tag{10.2}$$

For a fixed $\mathbf{g}$, directions of vectors $\mathbf{k}$ satisfying this equation determine two cones composed of reflected rays (Fig. 10.1). An intersection of the cones with the surface of a detector contributes to a diffraction pattern. See an example in Fig. 10.2.

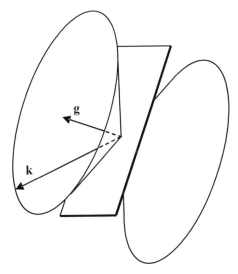

Fig. 10.1. Geometry of reflected beam for convergent incident beam.

Assuming that a number of points $\mathbf{x}^{(i)}$ ($i = 1, ..., I$) on the cones of one diffracting plane were detected, we can calculate the corresponding wave vectors $\mathbf{k}^{(i)} = \mathbf{x}^{(i)}/(\lambda(\mathbf{x}^{(i)} \cdot \mathbf{x}^{(i)})^{1/2})$. Since the vectors satisfy Eq.(10.2), there occurs

$$\pm \mathbf{d} \cdot \mathbf{k}^{(i)} = 1/2 \ , \tag{10.3}$$

where $\mathbf{d} = \mathbf{g}/g^2$. The sign in front of $\mathbf{k}^{(i)}$ must be chosen in a consistent way, i.e., all vectors $\mathbf{k}^{(i)}$ of the same cone must be given the same sign. The system

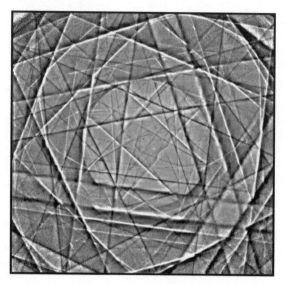

Fig. 10.2. Kikuchi electron diffraction pattern of Cu. (Courtesy E.Bouzy, Universite de Metz.) Because of small wavelength, the opening angles of the cones are large and the cones' sections are visible as straight lines.

of linear equations (10.3) can be solved with respect to **d**. The equations of the system are independent if the vectors $\mathbf{k}^{(i)}$ are not coplanar. Using the least squares method, the solution can be written as $\mathbf{d} = \pm X^{-1} \sum_i \mathbf{k}^{(i)}/2$, where the matrix X is defined by $X_{mn} = \sum_i k_m^{(i)} k_n^{(i)}$, and $k_m^{(i)}$ denotes the m-th component of $\mathbf{k}^{(i)}$. With known **d**, the sought vector **g** is given by $\mathbf{g} = \mathbf{d}/(\mathbf{d} \cdot \mathbf{d})$.

▣ **g** *from two points*

Various shapes of detectors can be devised but the most practical is a planar one. The intersection of the reflected rays with the plane of the detector leads to the diffraction pattern – a set of conics. If the opening angles of the cones are large (as in the case of Kikuchi patterns) the conics will appear in pairs. In this case, **g** can be determined in a simpler manner by using two particular points on the traces of the cones on the detector.

The equation of the detector's plane is given by $\mathbf{n} \cdot \mathbf{x} = L > 0$, where **x** represents points of the plane, **n** is a unit vector normal to the plane, and L is the detector-to-sample distance. The pattern center is determined by the direction of **n**; the appropriate point $\mathbf{x}^0$ on the detector is given by $\mathbf{x}^0 = L\mathbf{n}$. Let $\mathbf{x}^-$ and $\mathbf{x}^+$ denote the points closest to the pattern center on each of the two traces. Again, because $\mathbf{x}^\pm$ lie on the diffraction cones, the vectors $\mathbf{k}^\pm = \mathbf{x}^\pm/(\lambda(\mathbf{x}^\pm \cdot \mathbf{x}^\pm)^{1/2})$ satisfy the relation $\pm \mathbf{k}^\pm \cdot \mathbf{g} = g^2/2$. Another equation comes from the fact that $\mathbf{x}^\pm$, and thus $\mathbf{k}^\pm$, are chosen to be coplanar with **g**, i.e., $(\mathbf{k}^+ \times \mathbf{k}^-) \cdot \mathbf{g} = 0$. These three equations for **g** have a solution given by

$$\mathbf{g} = \mathbf{k}^+ - \mathbf{k}^- . \tag{10.4}$$

The plane normal opposite to the above one is obtained if the roles of the points $\mathbf{x}^-$ and $\mathbf{x}^+$ are interchanged, and thus, the sign of **g** is undetermined. ⊠

With a given diffraction pattern, one would like to 'solve it', i.e., to assign Miller indices to experimentally detected reflections. This crucial task is referred to as "indexing". For a known crystal structure it is relatively simple (but definitively not trivial). The solution is found by matching the vectors of the crystallographic reciprocal lattice and the vectors of the reciprocal lattice obtained from the experimental diffraction pattern. By applying Eq.(2.63) to the indices of the possible reflections, coordinates of the normals $\mathbf{h}$ are obtained from the crystallographic data. Because of crystal symmetry, there are families of symmetrically equivalent reflecting planes. Even if a crystal is not centrosymmetric, due to Friedel's law, patterns appear as if they came from a crystal with inversion symmetry; therefore, the symmetry operators of the crystal's Laue group must be applied to a given vector $\mathbf{h}$ to get all symmetrically equivalent members of the family. The vectors of the set of possible reflections will be denoted by $\mathbf{h}_m$, with $m = 1, ..., M$. Their coordinates are given in the crystal coordinate system. On the other hand, by analyzing a diffraction pattern we get from Eq.(10.1) the coordinates of normals $\mathbf{g}$ in the sample coordinate system. Let the number of them be N. These vectors will be denoted by $\mathbf{g}_n$, $(n = 1, ..., N)$.

Each legitimate vector $\mathbf{g}_n$ corresponds to a certain $\mathbf{h}_m$. The task is to match the figure composed of the vectors $\mathbf{g}_n$ with a congruent figure of a (yet to be determined) subset of $\mathbf{h}_m$ vectors (Fig. 10.3). This means that each element of the set $\{\mathbf{g}_n; n = 1, ..., N\}$ is matched to a unique element of $\{\mathbf{h}_m; m = 1, ..., M\}$. Formally, indexing is the procedure of determining injective mapping $\pi : \{1, ..., N\} \rightarrow \{1, ..., M\}$, i.e., to each of subscripts enumerating experimental vectors $\mathbf{g}_n$, π assigns one of the subscripts enumerating vectors $\mathbf{h}_m$. Symmetry operations map the set of $\mathbf{h}_m$ vectors onto itself and induce permutations of the indices $\{1, ..., M\}$. If σ is such a permutation, the assignment π is equivalent to $\sigma\pi$. Knowing one representative of the class of equivalent assignments, it is straightforward to get all other elements of that class.

Apart from experimental errors, there exists a rotation acting on vectors given in the sample coordinate system such that the transformed coordinates of a vector $\mathbf{g}_n$ are identical to coordinates of certain $\mathbf{h}_m$ in the crystal coordinate system. In other words, there exist an assignment π and an orthogonal matrix O such that $\mathbf{h}_{\pi(n)} = O\mathbf{g}_n$ for each $n = 1, ..., N$. Due to experimental errors, this relation is only approximately fulfilled. For a given pattern, the residue $\psi = \sum_{n=1}^{N}(\mathbf{h}_{\pi(n)} - O\mathbf{g}_n)^2$ provides quantitative measure of the misfit between experimental and crystallographic data. With the set of vectors $\mathbf{h}_m$ $(m = 1, ..., M)$ represented by a $3 \times M$ matrix H and vectors $\mathbf{g}_n$ $(n = 1, ..., N)$ represented by a $3 \times N$ matrix G, the relation $\mathbf{h}_{\pi(n)} = O\mathbf{g}_n$ can be written as

$$HP = OG , \qquad (10.5)$$

where P is a $M \times N$ match matrix given by $P_{mn} = 1$ if $\pi(n) = m$, $P_{mn} = 0$ otherwise; it satisfies $P^T P = I_N$ ($I_N - N \times N$ identity matrix). The residue ψ takes the form

$$\psi = \|HP - OG\|^2 . \qquad (10.6)$$

An assignment P and an orthogonal matrix O leading to the smallest possible value of ψ are considered to be solutions to the indexing problem and the

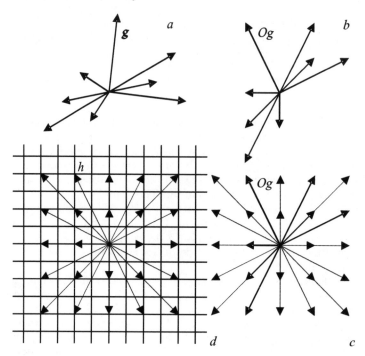

Fig. 10.3. Schematic explanation of the indexing procedure. *a*) A set of detected vectors $\mathbf{g}_n$. *b*) The rotated set ($O\mathbf{g}_n$). *c*) The rotated set and other vectors of the reciprocal lattice. *d*) The reciprocal lattice.

orientation determination problem, respectively. Since the diffraction pattern corresponds to a crystal point group with inversion center, the orientation is given by either O, if $\det(O) = +1$, or by $\Im O$ ($\Im$ – inversion), if $\det(O) = -1$. The search for a global minimum of $\psi = \psi(P, O)$ is a problem in the field of both combinatorial (P) and continuous (O) optimization. It can be seen as a kind of generalization of the Procrustes problem.

For a given P (i.e., with known assignment) the problem is reduced to determination of O by the minimization of $\psi_P = \psi(P, \cdot)$, i.e., to the classical Procrustes problem. The latter is solved with respect to O by the polar decomposition of HPG^T; cf. chapter 5. In particular, this concerns the case of manual indexing with the assignment of indices resolved by human analysis of the pattern.

In 'automatic' indexing of diffraction patterns, the assignment matrix P is unknown. One could enumerate all possible assignments of $\mathbf{h}_m$ to $\mathbf{g}_n$ vectors (i.e., all possible realizations of π and P) and after solving the Procrustes prolem for each case, take the match matrix leading to the smallest value of ψ as the solution of the indexing problem. Unfortunately, except the cases where M and N are small, the number of possible trials is prohibitive and this direct approach cannot be used.

Other forms of the residue
With the same angular mismatch, the larger magnitude of vectors $\mathbf{h}_{\pi(n)}$ and $\mathbf{g}_n$, the bigger their influence on the residue ψ. This effect does not occur if all $\mathbf{h}_m$ and $\mathbf{g}_n$ vectors are pre–normalized or if the residue is defined as $\Psi = \sum_{n=1}^{N} (\mathbf{h}_{\pi(n)} - O\mathbf{g}_n)^2 / \mathbf{g}_n^2$. With the matrix M given by $M = \mathrm{diag}\{1/\|\mathbf{g}_1\|, 1/\|\mathbf{g}_2\|, ..., 1/\|\mathbf{g}_N\|\}$, one has $\Psi = \|HPM - OGM\|^2$.

Alternatively to (10.6), the considered mixed continuous–combinatorial optimization problem can be approached using the residue

$$\sum_{m=1}^{M} \sum_{n=1}^{N} P_{mn} (\mathbf{h}_m - O\mathbf{g}_n)^2 \ , \tag{10.7}$$

which is linear with respect to P. ☒

Insight into complexity
From the relation (10.5), we get

$$P^T W_h P = W_g \ ,$$

where the symmetric (Gram) matrices $W_h = H^T H$ and $W_g = G^T G$ contain scalar products of $\mathbf{h}_m$ and $\mathbf{g}_n$ vectors, respectively. The matrix O is eliminated, and one faces the purely combinatorial problem of matching W_g with a sub-matrix of W_h by finding P which minimizes the squared norm of $P^T W_h P - W_g$

$$\|P^T W_h P - W_g\|^2 \ .$$

Here, the exponent of P entries is twice larger than in the case of ψ and the procedure is much more sensitive to experimental errors. But even if that sensitivity were acceptable, the problem is not easy. In the particular case of $M = N$, i.e., when P is a square matrix, the residue equals $const - 2\mathrm{tr}(P^T W_h P W_g)$. The combinatorial problem of determining the match P which leads to a global extremum of $\mathrm{tr}(P^T W_h P W_g)$ is called Quadratic Assignment Problem. (See, e.g., Pardalos, Rendl & Wolkowicz, 1994.) It is known to be $\mathcal{NP}$–hard, which in practical terms means that with increasing N the time required to solve the problem increases exponentially. Thus, the most reliable results would be expensive in terms of the computation time. ☒

Softassign matching
Analogous problems of matching are encountered in computerized imaging and are solved by genreally applicable methods. The mixed problem with its combinatorial part is transformed to purely continuous nonlinear problem. For instance, one of the algorithms uses the residue (10.7) with additional constraints. More precisely, it is based on the minimization of a function which for our purposes would have the form

$$\sum_{m=1}^{M} \sum_{n=1}^{N} \mathcal{P}_{mn} \left((\mathbf{h}_m - O\mathbf{g}_n)^2 - \alpha\right) + \sum_{m=1}^{M} \lambda_m^h \left(\left(\sum_{n=1}^{N+1} \mathcal{P}_{mn}\right) - 1\right)$$
$$+ \sum_{n=1}^{N} \lambda_n^g \left(\left(\sum_{m=1}^{M+1} \mathcal{P}_{mn}\right) - 1\right) + \beta^{-1} \sum_{m=1}^{M+1} \sum_{n=1}^{N+1} \mathcal{P}_{mn} \log \mathcal{P}_{mn} \ . \tag{10.8}$$

In the process of optimization, the $(M + 1) \times (N + 1)$ matrix $\mathcal{P}$ with positive real entries is expected to reach the integer form such that $\mathcal{P}_{mn} = P_{mn}$ for $m \leq M$ and $n \leq N$, $\mathcal{P}_{M+1\,n} = 1$ if $\mathbf{g}_n$ is spurious, $\mathcal{P}_{m\,N+1} = 1$ if $\mathbf{h}_m$ does not match any experimental reflection, and $\mathcal{P}_{mn} = 0$ in other cases. The Lagrange multipliers λ_m^h and λ_n^g are applied to ensure the constraints on the sums of row and column entries of $\mathcal{P}$. Moreover, a mechanism called barrier function is used; the entropy barrier function $\sum_{mn} \mathcal{P}_{mn} \log \mathcal{P}_{mn}$ enforces the positivity of the entries of $\mathcal{P}$. Positive α and β are control parameters. The algorithm is called 'softassign matching'. For details, see Rangarajan, Chui and Bookstein (1997). Schemes of this type are yet to be tested for indexing. ⊠

The number of trials can be considerably reduced by eliminating some of the $\mathbf{h}_m$ vectors. Scalar products of vectors are invariants of rotations. Thus, the scalar products between $\mathbf{g}_n$ vectors and the products between the matching $\mathbf{h}_{\pi(n)}$ vectors must agree to within the limits of experimental errors. Therefore, vectors $\mathbf{h}_m$ which do not fit into any of the configurations built of other vectors can be eliminated from the list of feasible assignments. This reduces the number of possible matrices P and the search for optimal O can be completed in a shorter time. Maximal errors of components of vectors $\mathbf{g}_n$, and subsequently, the error bounds of their scalar products, can be obtained from maximal errors of experimental parameters (i.e., parameters of the lines, camera length, pattern center, et cetera).

☐ *Combined uncertainity*
Generally, the "combined standard uncertainity" $\Delta(f)$ of $f = f(x_1, x_2, ..., x_k)$, resulting from the propagation of small errors of the parameters x_i, can be calculated from

$$\Delta(f) = \left(\sum_{i=1}^{k} \left(\frac{\partial f}{\partial x_i} \Delta(x_i) \right)^2 \right)^{1/2} ,$$

where $\Delta(x_i)$ is the "standard uncertainity" of the parameter x_i (e.g., Cruickshank, 1949, or a textbook on error analysis). ⊠

Ideally, the elimination process is implemented for all combinations of i–tuples of (not yet eliminated) $\mathbf{h}_m$ vectors, with $i = 1, 2, ..., N$. The criterion is simple: a given vector is eliminated if none of all possible i–tuples of $\mathbf{h}_m$ matching i–tuples of $\mathbf{g}_n$ contains that vector. Once certain vectors are eliminated, the procedure must be repeated for the same i because some vectors may have lost their partners and may be unable to fit into a configuration congruent to the experimental one. There are three possible outcomes of the above elimination process: 1. the assignment problem is uniquely solved, 2. the result is ambiguous, i.e., within specified error brackets a number of non-equivalent assignments are detected, 3. there is a contradiction, and no match can be found for certain vectors $\mathbf{g}_n$. The case 3 means that the error bounds on the experimental data are too restrictive, or there are spurious $\mathbf{g}_n$ vectors in the experimental data set. In the case of 2, the ambiguity can be resolved by selecting the solution minimizing ψ.

In order to handle experimental inaccuracies and spurious reflections in a reasonable time, an appropriate strategy is needed. The choice of strategy to search for a solution can be important for robustness and speed of indexing. Efficient strategies are usually based on dividing the job into smaller jobs by considering subsets of the complete set of available vectors $\mathbf{g}_n$. Since each subset may provide a different solution, there remains an issue of finding a compromise for indexing and for orientation. Another approach is to begin with narrower error bounds and increase them only if no solution was found. There is a chance that the solution will fit into the reduced limits, and the number of considered assignments will be much smaller. Otherwise, a contradiction will be detected and the bounds will have to be increased.

Even if there are no experimental errors, one may face ambiguities in indexing, which subsequently lead to ambiguities in orientation determination. In general, a solution is ambiguous if two non–equivalent sets of reciprocal vectors constitute congruent configurations. 'Non–equivalent' means that there is no crystal symmetry operation leading from one to the other.

The most obvious ambiguity arises from the Friedel's law. Although, in certain conditions the breakdown of the law is observed, in most cases it must be taken into account. Therefore, for crystals without the inversion center, the solution is ambiguous. In the case of crystals without improper symmetry operations, there is no distinction between patterns coming from the right and left enantiomorphic forms of the crystal. In the case of crystals with improper symmetries but without the inversion center, two non–equivalent orientations lead to identical patterns. These two orientations are mutually related. If the first one is given as a proper orthogonal matrix O, the second one can be expressed as $O^R O$, where O^R is the product of an improper symmetry operation of the crystal point group and the inversion. Generally, errors of similar nature may occur in the presence of the so–called diffraction enhancement of symmetry (see e.g. Sadanaga & Ohsumi, 1979).

Another ambiguity arises in the case with coplanar reciprocal vectors. There are two (possibly non–equivalent) orientations which lead to such configuration of vectors. The two orientations are mutually related. If the first one is given as a proper orthogonal matrix O, the second one can be expressed as $O^S O$, where O^S represents the rotation about the axis perpendicular to the plane of the vectors by the angle of 180°. If the axis happens to be of even foldness, the two solutions are symmetrically equivalent and there is no ambiguity.

Finally, if the true crystal symmetry is close to a higher symmetry (e.g., tetragonal with the ratio of unit cell parameters c/a close to 1) some patterns corresponding to non–equivalent orientations can be almost identical and this may lead to incorrect indexing.

10.3 Orientation Distributions from Pole Figures

Formally, an ideal pole figure $\mathsf{P_h}$ is defined based on the orientation distribution f by the equation

$$\mathsf{P_h(g)} = \frac{1}{4\pi} \int_{\{O(3)|\mathbf{h}=O\mathbf{g}\}} f(O)\mathrm{d}O \ , \tag{10.9}$$

where $\mathbf{g}$ and $\mathbf{h}$ are of equal magnitude and only the orientations O satisfying $\mathbf{h} = O\mathbf{g}$ contribute to the integral.

Let S_i and S'_j represent crystal and sample symmetry operations, respectively. With orientation distribution f satisfying the conditions $f(S_i O S_j'^{-1}) = f(O)$, the pole figure $\mathsf{P_h}$ has the symmetry

$$P_{S_i\mathbf{h}}(S_j'\mathbf{g}) = P_\mathbf{h}(\mathbf{g}) \ . \tag{10.10}$$

◰ *Symmetries of pole figures*

It follows from the sequence

$$P_{S_i\mathbf{h}}(S_j'\mathbf{g}) = \frac{1}{4\pi}\int_{\{O(3)|S_i\mathbf{h}=OS_j'\mathbf{g}\}} f(O)dO = \frac{1}{4\pi}\int_{\{O(3)|\mathbf{h}=S_i^{-1}OS_j'\mathbf{g}\}} f(O)dO$$

$$= \frac{1}{4\pi}\int_{\{O(3)|\mathbf{h}=O\mathbf{g}\}} f(S_iOS_j'^{-1})d(S_iOS_j'^{-1}) = \frac{1}{4\pi}\int_{\{O(3)|\mathbf{h}=O\mathbf{g}\}} f(O)dO = P_\mathbf{h}(\mathbf{g}) \ .$$

⊠

◰ *Inverse pole figures*

In analysis of textures, also the so-called inverse pole figures are used. The inverse pole figure $R_\mathbf{g}$ specified by a fixed vector $\mathbf{g}$ is defined as

$$R_\mathbf{g}(\mathbf{h}) = \frac{1}{4\pi}\int_{\{O(3)|\mathbf{h}=O\mathbf{g}\}} f(O)dO \ .$$

The inverse pole figures satisfy symmetry conditions analogous to (10.10) $R_{S_j'\mathbf{g}}(S_i\mathbf{h}) = R_\mathbf{g}(\mathbf{h})$. ⊠

When domains of orientation distributions are limited to $SO(3)$, the ideal pole figure is determined by

$$P_\mathbf{h}(\mathbf{g}) = \frac{1}{2\pi}\int_{\{SO(3)|\mathbf{h}=O\mathbf{g}\}} f(O)dO \ . \tag{10.11}$$

It is common in texture analysis to use this definition.

Pole figures can be measured experimentally. Usually, X–ray diffraction or neutron diffraction are used for that purpose. The source of radiation and the counter are aligned in such a way that $\mathbf{k}_0$ and $\mathbf{k}$ satisfy the Laue equation $\pm\mathbf{h} = \mathbf{k} - \mathbf{k}_0$ for a certain $\mathbf{h}$. Then, the sample is rotated through polar and azimuthal angles in steps, and for each step the intensity of reflected radiation is registered. That intensity (after some corrections) is proportional to the volume of the crystallites which happen to be in reflective positions. After normalization to 1, this is taken as the value of the pole figure at the direction of $\mathbf{g}$ in the sample coordinate system. Experimentally measured pole figure $P_\mathbf{h}^{exp}$ involves all symmetrically equivalent orientations of the crystal. Because of the Friedel's law, not only orientations O satisfying $+\mathbf{h} = O\mathbf{g}$ but also those complying with $-\mathbf{h} = O\mathbf{g}$ contribute to the integral. Thus,

$$P_\mathbf{h}^{exp}(\mathbf{g}) = \frac{1}{2N}\sum_i (P_{S_i\mathbf{h}}(\mathbf{g}) + P_{-S_i\mathbf{h}}(\mathbf{g})) \ , \tag{10.12}$$

where N is the number of symmetry operations. This relation with $P_\mathbf{h}$ given by (10.11) is pretentiously referred to as the "fundamental equation of texture analysis". Frequently, the experimental pole figures are incomplete and cannot be normalized; in such cases the sign of equality must be replaced by proportionality.

With known experimental pole figures $P_{\mathbf{h}}^{exp}$, (10.12) is an equation for f. However, since both $+\mathbf{h} = O\mathbf{g}$ and $-\mathbf{h} = O\mathbf{g}$ contribute to the integral, the equation has a non–trivial kernel, i.e., there exists a non–zero function, say $\tilde{\tilde{f}}$, on the orientation space such that $\int_{\{SO(3)|\mathbf{h}=\pm O\mathbf{g}\}} \tilde{\tilde{f}}(O)\mathrm{d}O = 0$ for all $\mathbf{h}$ and $\mathbf{g}$ (see, e.g., Matthies *et al.*, 1987). Generally, even with error free data, orientation distribution cannot be uniquely determined from pole figures of this type. The ambiguity appears in the form of false maxima (ghosts) on graphs of orientation distributions. Texture community calls the effect 'the ghost phenomenon'.

It is noteworthy that orientation (distribution) can be uniquely determined from a pole figure of a single crystal having the inversion center (or a small number of such crystals). Thus, the nature of the ambiguity of the ghost phenomenon is associated with the continuous character of the orientation density function. ⊠

Nevertheless, in practice close approximations of f can be calculated from small sets of pole figures (or even incomplete pole figures). For these numerical calculations a discretization is necessary. One type of discretization is by series expansion. Series expansion allows the kernel to be expressed in an explicit form.[2] Another approach is to tessellate the (asymmetric) domain of the orientation distribution into cells. The domains of pole figures usually inherit a tessellation from a discrete measurement. Orientation distribution and pole figures are approximated by piecewise constant functions. If p_i is a pole figure value at i-th cell of the pole figure domain, and f_j is a value of the orientation distribution at j-th cell of the orientation space, then the relation between them follows from (10.12) and can be written as

$$p_i \approx \sum_j A_{ij} f_j \ . \tag{10.13}$$

With given p_i, the above relation is a system of linear equations with respect to f_j. Additional conditions are $f_j \geq 0$ and $\sum_j f_j = 1$. As the kernel of (10.12) is nontrivial, the system is underdetermined and there is a range of solutions to (10.13).[3] Large part of them can be eliminated by the natural requirement that f_j must be non–negative. Additional more arbitrary criteria are used to single out a given solution. Some of them are associated with the method of solving the system. These are usually iteration procedures. For instance, there is a method which gives a solution with maximal entropy (5.15) which in the discrete form is given by $-\sum_i f_i \ln f_i$ (e.g., Schaeben, 1994).

Maximal entropy method
The algorithm of this method can be written as

---

[2] This subject is extensively discussed in books by Bunge (1982) and Matthies, Vinel, & Helming (1987).

[3] It must be remembered that experimental errors and inaccuracies caused by tessellations only blur the problem.

$$f_j^{(k+1)} = \prod_{i\ (A_{ij}\neq 0)} \left(\frac{p_i}{\sum_j A_{ij} f_j^{(k)}}\right)^{\lambda_k A_{ij}} ,$$

where the product is over all i for which $A_{ij} \neq 0$, $f_j^{(k)}$ is the value of f_j in the k–th iteration step, the initial values $f_j^{(0)}$ of f_j correspond to the random case $f(O) = 1$ and λ_k is a relaxation parameter. It is easy to understand 'the mechanism' of the procedure since the denominator contains $\sum_j A_{ij} f_j^{(k)}$, i.e., values of pole figures recalculated from the k–th approximation of the orientation distribution.

When the measured pole figures are incomplete, they are initially roughly normalized to the fraction of the sphere covered by the measurement. The normalization is corrected in each iteration step by using the normalization coefficient obtained from the recalculated pole figures. ⊠

The coefficients A_{ij} of the system (10.13).
The issue of calculating A_{ij} has a purely geometrical nature. The particular values depend on the parameterization and tessellations used. In the most common case when Euler angles are applied, the coefficients can be stored in a relatively compact form by the use of the fact that the pole figure projection depends on φ_1 only via $\varphi_1 - \beta$, where β is the azimuth angle on a pole figure. The coefficients can be stored for just one value of φ_1, say 0; when needed for another value of φ_1, these stored for 0 are suitably translated by β. This considerably affects efficiency of the program. ⊠

Integration paths expressed using matrices
The integration in pole figure calculation is performed over paths of orientations satisfying $\mathbf{h} = \pm O\mathbf{g}$. What is the form of the special orthogonal matrices O satisfying

$$\mathbf{h} = O\mathbf{g} \ ?$$

Without loosing generality, let $\mathbf{h}$ and $\mathbf{g}$ be normalized to unit magnitude. Moreover, let $\mathbf{h} \neq \pm\mathbf{g}$. There are two easy to guess solutions of $\mathbf{h} = O\mathbf{g}$. One is provided by Eq.(2.25), i.e. it is a rotation about $(\mathbf{h} + \mathbf{g})/\|\mathbf{h} + \mathbf{g}\|$ by π

$$O_1 = O((\mathbf{h} + \mathbf{g})/\sqrt{2(1 + \mathbf{h} \cdot \mathbf{g})}, \pi) \ .$$

The other one is a rotation about the axis perpendicular to both $\mathbf{h}$ and $\mathbf{g}$ by the angle between the two vectors

$$O_2 = O((\mathbf{h} \times \mathbf{g})/\sin\psi, \psi) \ , \quad \text{where} \quad \psi = \arccos(\mathbf{h} \cdot \mathbf{g})$$

Brief calculation shows that $O_1^T O_2 = O(\mathbf{g}, \pi)$. Based on Eq.(3.24), the geodesic line through O_1 and O_2 parameterized by ω has the form

$$O_1 O(\mathbf{g}, \omega) \ .$$

It easy to check that $O_1 O(\mathbf{g}, \omega)\mathbf{g}$ equals $\mathbf{h}$ for arbitrary ω. This is also true if $\mathbf{h} = \mathbf{g}$. In the case of $\mathbf{h} = -\mathbf{g}$, a rotation about an axis perpendicular to $\mathbf{h}$ by π is taken as O_1.

The line corresponding to $\mathbf{h} = -O\mathbf{g}$ is characterized in analogous way. ⊠

Integration paths on the unit quaternion sphere
The integration paths have interesting interpretation on the sphere of unit

quaternions. Let the quaternion q correspond to O, and let g and h be quaternions with zero scalar parts and vector parts given by normalized $\mathbf{g}$ and $\mathbf{h}$, respectively. From Eq.(2.49), the formula for the integration path in the quaternionic form is

$$h = \pm q \diamond g \diamond q^* \ .$$

Hence, $h^* \diamond q \diamond g = \pm q$. This is a (homogeneous) system of linear equations for the components of q. It can be written in the matrix form $Pq = \pm q$, where the quaternion is treated as a 4×1 matrix and $P = P(h, g)$ is a 4×4 matrix given by

$$P = \begin{bmatrix} P_{00} & P_{0k} \\ P_{l0} & P_{kl} \end{bmatrix} = \begin{bmatrix} g^i h^i & \varepsilon_{kij} g^i h^j \\ \varepsilon_{lij} g^i h^j & g^k h^l + g^l h^k - g^i h^i \delta_{kl} \end{bmatrix} \ .$$

The matrix P is symmetric and orthogonal. It has two doubly degenerate eigenvalues $+1$ and -1. Thus, the quaternions satisfying $h = +q \diamond g \diamond q^*$ lie on the intersection of the unit quaternion sphere and a two dimensional plane (through the origin) in the four dimensional quaternion space, i.e., on a great circle of the sphere. Those satisfying $h = -q \diamond g \diamond q^*$ constitute another circle. The scalar product of a quaternion q from the first circle and a quaternion q' from the second circle can be expressed as $q \cdot q' = (+Pq) \cdot (-Pq') = -q \cdot q'$. This means that $q \cdot q' = 0$ and any two vectors of the two circles are orthogonal (Schaeben, 1996). $\boxtimes$

11

Effective Elastic Properties of Polycrystals

T HE determination of effective elastic properties of polycrystalline materials is one of the example problems in application of crystallographic
textures. Its essence is to calculate the elastic constants of a polycrystal based
on single crystal data and on the texture and microstructure of the material.
The problem is a special case of the more general subject of effective physical properties of microinhomogeneous materials. In turn, this issue lies in the
domain of the theory of continuous media with stochastic properties. Inhomogeneity in single–phase polycrystals arises due to the anisotropy of crystallites
variously oriented in the material. Therefore, calculations of polycrystalline
effective constants involve averaging of tensor quantities over orientations.

Numerous methods of solving the problem of effective properties have been
proposed. Only some of them will be mentioned here. Moreover, our discussion
is limited to single phase statistically homogeneous polycrystals. Example
calculations are carried out for the special case of the quasi-isotropic aggregate
of crystallites with cubic crystal symmetry.

The problem of effective elastic constants has been presented in a number
of review articles. Best known among them are probably papers of Kröner
(e.g., 1971, 1976 (with Koch), 1986).

11.1 Definitions and Simplest Principles

We begin this chapter with a section recalling basics of elasticity theory. It will
establish the notation which will be used to define the effective constants. The
section will be closed with the description of the simplest averaging methods.

11.1.1 Elementary notions of elasticity

Let us begin with a brief reminder of the elementary linear elasticity theory.
The state of an elastically deformed material is determined by the strain tensor
$\epsilon = \{\epsilon_{ij}\}$. With small deformations, it is defined as the symmetrized gradient
of the displacement vector field $\mathbf{u}$

$$\epsilon_{ij} = \partial_{(i} u_{j)} \ . \tag{11.1}$$

For a given twice differentiable tensor field ϵ, the existence of a displacement **u** satisfying the above relation is guaranteed if

$$\varepsilon_{ijk}\varepsilon_{lmn}\partial_j\partial_m\epsilon_{kn} = 0 \ ;$$

these are the so–called compatibility conditions.

For small deformations the elastic energy density U is a quadratic function of the strain tensor components

$$U = c_{ijkl}\epsilon_{ij}\epsilon_{kl}/2 \ ,$$

where **c**, referred to as stiffness tensor, satisfies the following symmetry conditions

$$c_{ijkl} = c_{ijlk} = c_{klij} \ . \tag{11.2}$$

Stresses within the material are determined by the the symmetric (stress) tensor σ such that $\delta U = \sigma_{ij}\delta\epsilon_{ij}$. Its explicit linear relation to the strain tensor

$$\sigma_{ij} = c_{ijkl}\epsilon_{kl} \tag{11.3}$$

is called the Hooke's law. In the static case, the divergence of the stress tensor is balanced by the body forces **f**

$$\partial_j\sigma_{ij} + f_i = 0 \ . \tag{11.4}$$

Without body forces the equilibrium conditions have the form $\partial_j\sigma_{ij} = 0$.

The algebraic operations on shear, stress and stiffness tensors can be carried out equivalently using properly ascribed matrices. Symmetric 6×6 matrices are assigned to tensors with symmetry (11.2) and column 6×1 matrices are assigned to stress and strain–type tensors.

Matrices assigned to tensors
There exists a traditional Voigt assignment of a matrix to a tensor but it is not the same for all tensors. It is more convenient to use the following correspondence (see, e.g., Gubernatis & Krumhansl, 1975):
a) pairs of tensor subscripts 11, 22, 33, 23, 13 and 12 correspond to matrix subscripts 1,2,3,4,5 and 6, respectively,
b) if one of the matrix subscripts, either of a column or of a row, is greater than 3, then to obtain the matrix element, the corresponding entry of the tensor must be multiplied by $\sqrt{2}$,
c) if subscripts of both, column and row, are greater than 3, then to obtain the matrix element, the corresponding tensor entry should be multiplied by 2,
d) otherwise, the matrix entry is equal to the corresponding entry of the tensor.
Using the rules a), b) and d) the strain and stress–type tensors are ascribed to column 6×1 matrices. ⊠

Since U is positive for any non-zero strain, the matrix corresponding to **c** is positive definite. Therefore, it is non–singular and, subsequently, there exist the inverse matrix and the corresponding (compliance) tensor **s** such that

$$\epsilon_{ij} = s_{ijkl}\sigma_{kl} \ .$$

This is another form of the Hooke's law. Symmetry properties of the compliance tensor are analogous to (11.2).

11.1.2 Notation

For brevity, the following notation will be used. The symbol T^2 denotes the set of second rank symmetric tensors, i.e., for ζ in T^2 there occurs $\zeta_{ij} = \zeta_{ji}$. The strain and stress tensors belong to T^2. The set of fourth rank tensors satisfying $t_{ijkl} = t_{jikl} = t_{ijlk}$ is denoted by $\overline{\mathsf{T}}^4$. Moreover, T^4 is the set of fourth rank tensors with $t_{ijkl} = t_{jikl} = t_{klij}$. Thus, both the stiffness and the compliance tensors belong to T^4. Obviously, T^4 is a subset of $\overline{\mathsf{T}}^4$. For $\mathbf{t}$ in $\overline{\mathsf{T}}^4$ and ζ belonging to T^2, the short notation $\mathbf{t}\zeta$ denotes the tensor with components $t_{ijkl}\zeta_{kl}$. Moreover, for $\mathbf{t}^1, \mathbf{t}^2$ in $\overline{\mathsf{T}}^4$, the notation $\mathbf{t}^1\mathbf{t}^2$ denotes $t^1_{ijkl}t^2_{klmn}$. The tensor $\mathbf{I}$ defined by

$$I_{ijkl} = (\delta_{ik}\delta_{jl} + \delta_{il}\delta_{jk})/2$$

serves as the identity of the algebra $\overline{\mathsf{T}}^4$. In particular, for the stiffness and compliance tensors, $\mathbf{cs} = \mathbf{I} = \mathbf{sc}$. Taking this into account, the notation $\mathbf{s} = \mathbf{c}^{-1}$ makes sense.

The number of independent tensor components of $\mathbf{c}$ and $\mathbf{s}$ is limited if the material exhibits a symmetry (e.g., Nye, 1957). Tensor $\mathbf{c} \in \mathsf{T}^4$ describing material properties of a crystal from the cubic crystallographic system is determined by three independent parameters; it is convenient to use

$$\kappa = c_{1111} + 2c_{1122} , \quad \rho = c_{1111} - c_{1122} , \quad \gamma = c_{1212} .$$

In this case, the notation

$$\mathbf{c} = \mathbf{cub}(\kappa, \rho, \gamma)$$

is used. Similarly, isotropic tensors (satisfying $\rho = \gamma$) will be denoted by the symbol

$$\mathbf{iso}(\kappa, \gamma) := \mathbf{cub}(\kappa, \gamma, \gamma) .$$

Besides the fact that linear operations on the $\mathbf{cub}$ tensors are linear operations on its parameters κ, ρ and γ, there occurs the following very convenient property: for $\mathbf{t}^1 = \mathbf{cub}(\kappa_1, \rho_1, \gamma_1)$ and $\mathbf{t}^2 = \mathbf{cub}(\kappa_2, \rho_2, \gamma_2)$, one has

$$\mathbf{t}^1\mathbf{t}^2 = \mathbf{cub}(\kappa_1\kappa_2, \rho_1\rho_2, \gamma_1\gamma_2)$$

(Walpole, 1966). It considerably simplifies calculations for the cubic and isotropic cases.

11.1.3 Averages

Ensemble average of a given quantity at a given point of the sample is simply the average over all potential values the quantity may take at that point. For example, the ensemble average $<\mathbf{c}>(\mathbf{x})$ of $\mathbf{c}$ at the point for $\mathbf{x}$ is obtained by averaging all stiffness tensors which could be possibly assigned to $\mathbf{x}$ (with all tensors given in the same coordinate system).

It is usually assumed that the sample is statistically homogeneous. The notion of statistical homogeneity is intuitively obvious: the sample is considered to be statistically homogeneous with respect to elastic properties if the ensemble average $<\mathbf{c}>$ does not depend on the location.

Let us consider a region V large in comparison to the size of crystallites. The notation will be slightly abused by using the same symbol V to denote the volume of the region, and by denoting by $<\mathbf{c}>$ averages other than the ensemble average. The volume average of $\mathbf{c}$ over V is defined by

$$<\mathbf{c}>= \frac{1}{V} \int_V \mathbf{c}(\mathbf{x}) \, \mathrm{d}_3\mathbf{x} \ ,$$

where $\mathbf{c}(\mathbf{x})$ represents the stiffness tensor at the point $\mathbf{x}$ of the material.

The elastic properties at a given point are determined by the orientation O of the crystallite containing that point. In other words, the tensor field $\mathbf{c}$ is actually determined on the orientation space ($\mathbf{c} = \mathbf{c}(O)$) and through that – on the sample. The orientation average can be written as

$$<\mathbf{c}>= \int_{SO(3)} \mathbf{c}(O) f(O) \mathrm{d}O \ ,$$

where $f(O)$ is the orientation distribution.[1]

11.1.4 Effective constants

From the macroscopic point of view, the statistically homogeneous material appears as homogeneous. Therefore, the existence of a constant tensor $\mathbf{c}^*$ describing the overall elastic properties is expected. The elastic energy of the statistically homogeneous aggregate is assumed to be expressible as a quadratic function of macro–strain and, moreover, the macro–strain should be related to the macro–stress through macroscopic Hooke's law. In both cases, the tensor $\mathbf{c}^*$ is expected to provide appropriate coefficients.

The question is whether the macro-variables can be expressed through local variables. Due to the statistical homogeneity, the macro-variables are constant provided the boundary conditions are homogeneous. The standard procedure of defining the effective tensors uses the homogeneous conditions

$$u_i \mid_{\partial V} = \bar{\epsilon}_{ij} x_j \ , \tag{11.5}$$

or

$$\sigma_{ij} \mid_{\partial V} n_j = \bar{\sigma}_{ij} n_j \ , \tag{11.6}$$

where $\bar{\epsilon}$ and $\bar{\sigma}$ are constant tensors. The volume averages $<\epsilon>$ and $<\sigma>$ are equal to $\bar{\epsilon}$ and $\bar{\sigma}$, respectively.

One has

$$<\epsilon>_{ij}= \frac{1}{V} \int_V \partial_{(i} u_{j)} \mathrm{d}_3\mathbf{x} = \frac{1}{V} \int_{\partial V} u_{(i} n_{j)} \mathrm{d}_2\mathbf{x} = \frac{1}{V}\bar{\epsilon}_{k(i} \int_{\partial V} n_{j)} x_k \mathrm{d}_2\mathbf{x} = \bar{\epsilon}_{ij} \ ,$$

$$<\sigma>_{ij}= \frac{1}{V} \int_V \sigma_{ij} \mathrm{d}_3\mathbf{x} = \frac{1}{V} \int_V \left(\partial_k(\sigma_{ik} x_j) - (\partial_k \sigma_{ik}) x_j \right) \mathrm{d}_3\mathbf{x} = \frac{1}{V} \int_{\partial V} \sigma_{ik} x_j n_k \mathrm{d}_2\mathbf{x} = \bar{\sigma}_{ij} \ ;$$

---

[1] It is assumed throughout this chapter that the crystal has the center of symmetry, and the integration is limited to $SO(3)$.

Gauss theorem and the equilibrium conditions were used above. $\boxtimes$

With $<\epsilon>=\bar{\epsilon}$ and $<\sigma>=\bar{\sigma}$ as macro–variables the average energy density $<U>$ and the Hooke's law have the form

$$<U>=<\epsilon>\ \mathbf{c}^*\ <\epsilon>\ /2 \quad \text{and} \quad <\sigma>=\mathbf{c}^*\ <\epsilon>\ . \tag{11.7}$$

For consistency, the Hill's condition

$$<\epsilon\sigma>=<\epsilon><\sigma>$$

must be fulfilled. It follows from $<\epsilon\sigma>=<\epsilon\mathbf{c}\epsilon>=\ 2\ <U>=<\epsilon>\ \mathbf{c}^*\ <\epsilon>=$ $<\epsilon><\sigma>$. That requirement is satisfied for homogeneous boundary conditions (Bishop & Hill, 1951).

This is shown using the sequence

$$<\epsilon\sigma>= \frac{1}{V}\int_V \sigma_{ij}\partial_j u_i \mathrm{d}_3\mathbf{x} = \frac{1}{V}\int_{\partial V} \sigma_{ij}u_i n_j \mathrm{d}_2\mathbf{x}$$

$$= \begin{cases} (1/V)\bar{\epsilon}_{ik}\int_{\partial V}\sigma_{ij}x_k n_j \mathrm{d}_2\mathbf{x} = (1/V)\bar{\epsilon}_{ij}\int_V \sigma_{ij}\mathrm{d}_3\mathbf{x} \\ (1/V)\bar{\sigma}_{ij}\int_{\partial V}u_i n_j \mathrm{d}_2\mathbf{x} \quad = (1/V)\bar{\sigma}_{ij}\int_V \epsilon_{ij}\mathrm{d}_3\mathbf{x} \end{cases} =<\epsilon><\sigma>\ ;$$

the upper and lower lines correspond to the boundary conditions of the first and second kind, respectively. Again, the Gauss theorem and the conditions of local equilibrium were used. $\boxtimes$

The relations (11.7) serve as equivalent definitions of the effective constants $\mathbf{c}^*$. The defining relationship is sometimes written in the form

$$\mathbf{c}^*\ <\epsilon>=<\mathbf{c}\epsilon> \quad \text{or} \quad <\epsilon>\ \mathbf{c}^*\ <\epsilon>=<\epsilon\mathbf{c}\epsilon>\ , \tag{11.8}$$

Alternatively, the effective compliance tensor $\mathbf{s}^* = (\mathbf{c}^*)^{-1}$ can be defined as

$$\mathbf{s}^*\ <\sigma>=<\mathbf{s}\sigma> \quad \text{and} \quad <\sigma>\ \mathbf{s}^*\ <\sigma>=<\sigma\mathbf{s}\sigma>\ . \tag{11.9}$$

A correct procedure for the determination of effective constants is expected to provide the same result no matter which of the relations was the starting point of the calculations. For a simple tensor averaging, it means that it gives mutually reciprocal results when applied to mutually reciprocal tensors $\mathbf{c}$ and $\mathbf{s}$; in other words, the operation of averaging should commute with the matrix inversion.

11.1.5 Voigt and Reuss methods

The approaches of Voigt (1889, 1928) and Reuss (1929) are not only chronologically the first methods to determine the effective elastic constants but also, for reasons explained below, they are of fundamental importance for the whole problem. In the Voigt method, the average stiffness tensor is taken as the tensor describing the effective properties

$$\mathbf{c}^* =<\mathbf{c}>\ .$$

That would be true with (11.8) if the strains were homogeneous throughout the material: for $\epsilon(\mathbf{x}) = \text{const} = <\epsilon>$, one has $\mathbf{c}^* <\epsilon> = <\mathbf{c}\epsilon> = <\mathbf{c}><\epsilon>$, and because $<\epsilon>$ is arbitrary then $\mathbf{c}^* = <\mathbf{c}>$.

The Reuss method is dual to that by Voigt. The average of the compliance tensor is taken as the determinant of the effective properties $\mathbf{s}^* = <\mathbf{s}>$. In terms of the stiffness tensor this can be equivalently written as

$$\mathbf{c}^* = <\mathbf{c}^{-1}>^{-1} \ .$$

The Reuss average would be obtained from the definition (11.9) of the effective constants if the stresses were homogeneous ($\sigma(\mathbf{x}) = \text{const}$),

The Voigt and Reuss methods are based on the same average but apply different versions of the definition of the effective constants. In general, the results they give are different. This follows from the simple fact that arithmetic and harmonic means are different. Sometimes, the tensors $\mathbf{c}^V = <\mathbf{c}>$ and $\mathbf{c}^R = <\mathbf{c}^{-1}>^{-1}$ are called Voigt and Reuss averages of $\mathbf{c}$, respectively. The Voigt and Reuss averages are crucial for the issue of effective properties because, in certain sense, they bound the correct result. This follows from the variational principles of elasticity.

11.1.6 Voigt–Reuss bounds

Among all piecewise differentiable displacements which fulfill given boundary conditions, those satisfying the equilibrium equations correspond to a minimum of potential energy. All we are going to use is a version of this principle limited to the homogeneous boundary conditions: Let the continuous and piecewise differentiable displacements $\mathbf{u}$ and $\mathbf{u}^\star$, both satisfying the homogeneous boundary conditions (11.5), be given on V. Moreover, let $\epsilon_{ij} = \partial_{(i}u_{j)}$ and $\epsilon_{ij}^\star = \partial_{(i}u_{j)}^\star$. If $\mathbf{c}\epsilon$ satisfies the equilibrium relations, then

$$<\epsilon\mathbf{c}\epsilon> \ \leq \ <\epsilon^\star\mathbf{c}\epsilon^\star> \ .$$

Similar principles formulated for complementary energy and the second boundary value problem provide other inequalities of that type

$$<2\sigma <\epsilon> -\sigma\mathbf{s}\sigma> \ \geq \ <2\sigma^\star <\epsilon> -\sigma^\star\mathbf{s}\sigma^\star> \ ,$$

$$<\sigma\mathbf{s}\sigma> \ \leq \ <\sigma^\star\mathbf{s}\sigma^\star> \ ,$$

$$<2\epsilon <\sigma> -\epsilon\mathbf{c}\epsilon> \ \geq \ <2\epsilon^\star <\sigma> -\epsilon^\star\mathbf{c}\epsilon^\star> \ ,$$

where σ, ϵ and $\sigma^\star$, $\epsilon^\star$ are the actual and the trial fields, respectively.

With the trial field $\epsilon^\star$ given by $\epsilon^\star = <\epsilon>$, the first inequality leads to $<\epsilon> \mathbf{c}^* <\epsilon> \ \leq \ <\epsilon><\mathbf{c}><\epsilon>$ and this is true for arbitrary $<\epsilon>$. By convention, this is shortly written as $\mathbf{c}^* \ \leq \ <\mathbf{c}> =: \mathbf{c}^V$.[2] On the other hand, the second inequality and $\sigma^\star = <\sigma>$ give $\mathbf{s}^* \ \leq \ <\mathbf{s}>$ which is equivalent to $\mathbf{c}^* \ \geq \ <\mathbf{c}^{-1}>^{-1} =: \mathbf{c}^R$. Putting both results together, one finally has

$$\mathbf{c}^R \ \leq \ \mathbf{c}^* \ \leq \ \mathbf{c}^V \ .$$

Thus, Voigt and Reuss averages are upper and lower bounds for the tensor of effective constants (Hill, 1952). The remaining two inequalities give the same result.

---

[2] In general, $\mathbf{t}_1 < \mathbf{t}_2$ or $\mathbf{t}_1 \leq \mathbf{t}_2$ mean that the matrix corresponding to $\mathbf{t}_2 - \mathbf{t}_1$ is positive definite or positive semi–definite, respectively.

11.1.7 Orientation average *continued*

Let the tensor **t** have the symmetries of T^4. Let $\underline{t}_{ijkl}$ be components of that tensor in the coordinate system related to the crystal lattice of a crystallite and let O_{ij} be entries of the orthogonal matrix corresponding to the crystallite's orientation. Following the convention used in textures, the components of $\mathbf{t} = \mathbf{t}(O)$ in the laboratory (sample) coordinate system are given by

$$t_{ijkl} = O_{si}O_{tj}O_{uk}O_{wl}\underline{t}_{stuw} \; .$$

Assuming that f denotes orientation distribution, the symbol $\mathcal{M}^f(\mathbf{t})$ denotes the orientation average of **t**,

$$\mathcal{M}^f(\mathbf{t}) = \int_{SO(3)} f(O)\mathbf{t}(O)\,\mathrm{d}O \; .$$

In the explicit notation, the above formula can be written as $\left(\mathcal{M}^f(\mathbf{t})\right)_{ijkl} = M^f_{ijkl\,stuw}\underline{t}_{stuw}$, where $M^f_{ijkl\,stuw} = \int_{SO(3)} O_{si}O_{tj}O_{uk}O_{wl}f(O)\mathrm{d}O$. (Ganster & Geiss (1985) describe a method of calculating the integrals $M^f_{ijkl\,stuw}$.) Now, the determination of Voigt and Reuss averages is reduced to the orientation averages and one can write $\mathbf{c}^V = \mathcal{M}^f(\mathbf{c})$ and $\mathbf{s}^R = \mathcal{M}^f(\mathbf{s})$.

There are two independent conditions provided by the rotation invariants of tensors (e.g., Leibfried, 1953); because $M^f_{iijj\,stuw} = \delta_{st}\delta_{uw}$ and $M^f_{ijij\,stuw} = \delta_{su}\delta_{tw}$,

$$\left(\mathcal{M}^f(\mathbf{t})\right)_{iijj} = \underline{t}_{iijj} \quad \text{and} \quad \left(\mathcal{M}^f(\mathbf{t})\right)_{ijij} = \underline{t}_{ijij} \; . \tag{11.10}$$

On the other hand, when the orientation distribution is uniform ($f(O) = 1$) the average tensor is isotropic and determined by two independent variables; with known $\underline{t}_{ijkl}$, the relations (11.10) fully determine the isotropic tensor $\mathcal{M}^{f=1}(\mathbf{t})$. E.g., it is straightforward to check that

$$\mathcal{M}^{f=1}\left(\mathbf{cub}(\kappa, \rho, \gamma)\right) = \mathbf{iso}(\kappa, (2\rho + 3\gamma)/5) \; . \tag{11.11}$$

Further on, for the case with uniform orientation distribution the symbol $\mathcal{M} = \mathcal{M}^{f=1}$ will be used.

Geometric interpretation of Voigt and Reuss averages
The average of tensors over orientations described above has the following geometric property: Let the 'scalar' product $\mathbf{t}^1 \cdot \mathbf{t}^2$ of tensors $\mathbf{t}^1$ and $\mathbf{t}^2$ from T^4 be given by

$$\mathbf{t}^1 \cdot \mathbf{t}^2 = t^1_{ijkl}t^2_{klij} \; .$$

The norm and the distance d can be defined in the standard way

$$d(\mathbf{t}^1, \mathbf{t}^2) = \|\mathbf{t}^1 - \mathbf{t}^2\| \; , \quad \text{where} \quad \|\mathbf{t}\| = \sqrt{\mathbf{t} \cdot \mathbf{t}}$$

(Rychlewski, 1984; Krause, Kuska & Wedell, 1989). Isotropic tensors preserve their form when rotated, i.e., isotropic tensors constitute a linear subspace invariant under rotations. The complete space is the direct sum of the subspace of isotropic tensors and its orthogonal complement: an arbitrary **t** can be uniquely decomposed into

$\mathbf{t} = \mathbf{t}^{isotrop} + \mathbf{t}^{orthog}$, where $\mathbf{t}^{isotrop}$ is the isotropic part and $\mathbf{t}^{orthog}$ satisfies the orthogonality relation $\mathbf{t}^{isotrop} \cdot \mathbf{t}^{orthog} = 0$. With $\mathbf{x}$ being a varying isotropic tensor, the distance $d(\mathbf{t}, \mathbf{x})$, takes absolute minimum at $\mathbf{x} = \mathbf{t}^{isotrop}$. For the uniform orientation distribution, the average $\mathcal{M}(\mathbf{t})$ is the isotropic part of $\mathbf{t}$: $\mathcal{M}(\mathbf{t}) = \mathbf{t}^{isotrop}$. Going back to the effective properties, for $f(O) = 1$ the Voigt average is equal to the isotropic part of the stiffness tensor $\mathbf{c}$. Analogously, Reuss average is the isotropic part of the tensor $\mathbf{s}$. More complete discussion of the subject was given by Gazis, Tadjbakhsh & Toupin (1963). ⊠

⊡ *Invariance of the bulk modulus*

The invariance of κ in (11.11) is a special case of a more general property shown by Walpole (1985). In the case of cubic crystal symmetry the bulk modulus is invariant of the general class of solutions constructed according to the following scheme: a) local strain is expressed as $\epsilon = \mathbf{A} <\epsilon>$, where $\mathbf{A}$ is an operator satisfying $<\mathbf{A}> = \mathbf{I}$, b) from the definition $\mathbf{c}^* <\epsilon> = <\mathbf{c}\epsilon>$ and due to arbitrariness of $<\epsilon>$, one has $\mathbf{c}^* = <\mathbf{cA}>$. Both Voigt and Reuss procedures belong to that class. In the case of Voigt method $\mathbf{A} = \mathbf{I}$, and $\mathbf{A} = \mathbf{s} <\mathbf{s}>^{-1}$ leads to the Reuss average.

The modulus κ (related to the standard bulk modulus K via the relation $\kappa = 3K$) is defined by $\kappa = c_{iijj}/3$. In the case of cubic symmetry, $\kappa \Delta_{ijmn} = \Delta_{ijkl} c_{klmn}$, where $\Delta_{ijkl} = \delta_{ij}\delta_{kl}$. There occurs $\kappa^* \Delta = \Delta \mathbf{c}^* = \Delta <\mathbf{cA}> = <(\Delta\mathbf{c})\mathbf{A}> = <\kappa\Delta\mathbf{A}>$. The quantity $\kappa\Delta$, as constant, can be taken outside the brackets $<\cdot>$, i.e., $\kappa^*\Delta = \kappa\Delta <\mathbf{A}> = \kappa\Delta$. Hence $\kappa^* = \kappa$. ⊠

11.1.8 Hill average

Hill method (1952) in its original formulation concerns quasi-isotropic materials. When generalized to the anisotropic case, it has the form

$$\mathbf{c}^* = (\mathbf{c}^V + \mathbf{c}^R)/2 . \tag{11.12}$$

This arithmetic average of $\mathbf{c}^V$ and $\mathbf{c}^R$ fulfills the inequality $\mathbf{c}^R \leq \mathbf{c}^* \leq \mathbf{c}^V$. On the other hand, we can follow the concept of Shukla and Padial (1973) and take the 'harmonic' average of bounds as the effective tensor

$$\mathbf{s}^* = (\mathbf{s}^V + \mathbf{s}^R)/2 \quad \text{or equivalently} \quad (\mathbf{c}^*)^{-1} = \left((\mathbf{c}^V)^{-1} + (\mathbf{c}^R)^{-1}\right)/2 .$$

Again, the result will be within Voigt–Reuss bounds, but in general it will be different than (11.12).

The requirement that averaging of $\mathbf{c}$ and $\mathbf{s}$ tensors gives mutually inverse results is satisfied by a quantity defined in the form of a series. Let

$$\mathbf{c}^{n+1} = (\mathbf{c}^n + \mathbf{c}_n)/2 \quad \text{and} \quad (\mathbf{c}_{n+1})^{-1} = \left((\mathbf{c}^n)^{-1} + (\mathbf{c}_n)^{-1}\right)/2 ,$$

where $\mathbf{c}^0 = \mathbf{c}^V$ and $\mathbf{c}_0 = \mathbf{c}^R$. The common limit at $n \to \infty$ is considered to give the effective constants

$$\mathbf{c}^* = \lim_{n\to\infty} \mathbf{c}^n = \lim_{n\to\infty} \mathbf{c}_n .$$

With analogous procedure applied to get the effective compliance tensor $\mathbf{s}^*$ there occurs $\mathbf{c}^*\mathbf{s}^* = \mathbf{I} = \mathbf{s}^*\mathbf{c}^*$. However, that is not the only procedure satisfying this requirement. For example, the 'geometric' method (Morawiec, 1989) is another one.

Geometric average

The procedure is based on the exponential and logarithmic mappings for tensors with symmetries (11.2). Because of the relation of such tensors to symmetric positive definite matrices, the mappings described in section 3.2 can be applied without any modifications. The exponential mapping is given by $\mathbf{exp}(\mathbf{t}) = \sum_{n=0}^{\infty} \mathbf{t}^n / n!$. There is a neighborhood of the unity $\mathbf{I}$, where the series $\mathbf{log}(\mathbf{t}) = -\sum_{n=1}^{\infty} (\mathbf{I} - \mathbf{t})^n / n$ is convergent so the **log** mapping is correctly defined in that neighborhood. The effective tensor $\mathbf{c}^*$ is considered to be given by

$$\mathbf{c}^* = \mathbf{exp} <\mathbf{log}(\mathbf{c})> \ .$$

The requirement that averaging of mutually inverse tensors gives mutually inverse results is satisfied because $\mathbf{exp} <\mathbf{log}(\mathbf{c}^{-1})> = (\mathbf{exp} <\mathbf{log}(\mathbf{c})>)^{-1}$. The condition that the argument of the **log** mapping has to be in the neighborhood of $\mathbf{I}$ is not a limitation; due to the homogeneity of the operation $(\mathbf{exp} <\mathbf{log}(\alpha\mathbf{c})> = \alpha\mathbf{exp} <\mathbf{log}(\mathbf{c})>$ $(\alpha > 0)$, the positive definite stiffness tensor can be brought to the neighborhood of $\mathbf{I}$ by multiplying it by a non-negative number. ⊠

11.2 Perturbation Methods

The ad hoc Hill and 'geometric' procedures lack physical justification. There are numerous other methods of similar type with some arbitrary assumptions and the information about the structure of the material reduced to orientation distribution; see (Morawiec, 1994) for a review. Differently than the simple deterministic approaches, the statistical methods have more rigorous theoretical basis; essentially, they are solutions of the equilibrium equation. However, in order to go beyond Voigt–Reuss approximation, these more sophisticated methods require not only orientation distribution, but also statistical information about the structure of the material which includes correlations between orientations of spatially separated points. The important feature of these methods is that the agreement of the result they give with the actual value is adequate to the information used in calculation; to be more specific, the agreement depends on the order of correlation functions of the stiffness tensor field $\mathbf{c}$. In principle, an exact solution can be obtained provided full information about the sample is known. In reality, the information is limited. Therefore, the essence of the problem is to determine the optimal bounds for a given portion of available information.

This section illustrates mechanisms of such sophisticated approaches. It also contains examples for trivial textures and microstructures for which analytical calculations can be carried out.

11.2.1 Expression for $\varepsilon(\mathbf{x})$

As was already mentioned, the approach we are going to describe is based on the equilibrium equation. The derivation of formulas used in this section is relatively simple when the general expression for the solution of the equilibrium equation is taken into account. Such solution can be explicitly expressed with the help of the Green's function formalism. Necessary information is gathered below.

☐ *Solutions of equations of elastostatics*

We will briefly review essentials concerning application of Green's function in elastostatics. Based on (11.1) and (11.3), the equilibrium equation (11.4) takes the form

$$\partial_j \left(c_{ijkl}(\mathbf{x}) \partial_k u_l(\mathbf{x}) \right) = -f_i(\mathbf{x}) \ , \quad \mathbf{x} \in V \ , \tag{11.13}$$

where the displacement $\mathbf{u}$ is assumed to be of class C^1, and body forces $\mathbf{f}$ are supposed to be continuous on V. Let the boundary conditions be of the form

$$u_i(\mathbf{x}) = \varphi_i(\mathbf{x}) \ , \quad \mathbf{x} \in \partial V \quad \text{(first kind boundary conditions)}$$

or

$$n_j c_{ijkl}(\mathbf{x}) \partial_k u_l(\mathbf{x}) = \psi_i(\mathbf{x}) \ , \quad \mathbf{x} \in \partial V \quad \text{(second kind boundary conditions)} \ ,$$

where φ_i and ψ_i are known functions on the boundary ∂V. We are concerned with the tensor function $\mathbf{G}$ determined on $V \times V$ and being the solution of the equilibrium equation for the unit force $\delta_{im} \delta(\mathbf{x} - \mathbf{x}')$ applied at the point $\mathbf{x}'$ and directed along axes of the coordinate system

$$\partial_j \left(c_{ijkl}(\mathbf{x}) \partial_k G_{ml}(\mathbf{x}', \mathbf{x}) \right) = -\delta_{im} \delta(\mathbf{x} - \mathbf{x}') \ . \tag{11.14}$$

Using the Green's function, the solution of Eq.(11.13) can be expressed as[3]

$$u_i(\mathbf{x}) = \int_V G_{ij}(\mathbf{x}, \mathbf{x}') f_j(\mathbf{x}') d_3 \mathbf{x}'$$
$$- \int_{\partial V} [u_j(\mathbf{x}') c_{ljkm}(\mathbf{x}') \partial_k' G_{im}(\mathbf{x}, \mathbf{x}') - G_{ij}(\mathbf{x}, \mathbf{x}') c_{ljmk}(\mathbf{x}') \partial_m' u_k(\mathbf{x}')] n_l d_2 \mathbf{x}' \ . \tag{11.16}$$

Equation (11.16) involves both φ as well as ψ; on the other hand, each of these functions separately determines the solution of (11.13). By establishing the boundary conditions for $\mathbf{G}$ in Eq.(11.14) as $G_{ij}(\mathbf{x}, \mathbf{x}') = 0$, for $\mathbf{x}' \in \partial V$, the formal solution to the problem of the first kind is obtained

$$u_i(\mathbf{x}) = \int_V G_{ij}(\mathbf{x}, \mathbf{x}') f_j(\mathbf{x}') d_3 \mathbf{x}' - \int_{\partial V} n_l \varphi_j(\mathbf{x}') c_{ljkm}(\mathbf{x}') \partial_k' G_{im}(\mathbf{x}, \mathbf{x}') d_2 \mathbf{x}' \ . \tag{11.17}$$

Similarly, if tensor $\mathbf{G}$ satisfies (11.14) and $n_l c_{ljkm}(\mathbf{x}') \partial_k' G_{im}(\mathbf{x}, \mathbf{x}') = 0$ for $\mathbf{x}' \in \partial V$, then

$$u_i(\mathbf{x}) = \int_V G_{ij}(\mathbf{x}, \mathbf{x}') f_j(\mathbf{x}') d_3 \mathbf{x}' + \int_{\partial V} G_{ij}(\mathbf{x}, \mathbf{x}') \psi_j(\mathbf{x}') d_2 \mathbf{x}' \ . \tag{11.18}$$

In the case of unbounded (infinite) medium with $V = V^\infty$ and vanishing φ and ψ, the displacement $u_i(\mathbf{x})$ is given by

$$u_i(\mathbf{x}) = \int_V G_{ij}(\mathbf{x}, \mathbf{x}') f_j(\mathbf{x}') d_3 \mathbf{x}' \ . \tag{11.19}$$

---

[3] To show this, one needs to apply the Gauss theorem to the integral over ∂V and use the relation

$$\partial_l' [u_j(\mathbf{x}') c_{ljkm}(\mathbf{x}') \partial_k' G_{im}(\mathbf{x}, \mathbf{x}') - G_{ij}(\mathbf{x}, \mathbf{x}') c_{ljmk}(\mathbf{x}') \partial_m' u_k(\mathbf{x}')]$$
$$= u_j(\mathbf{x}') \partial_l' \left(c_{ljkm}(\mathbf{x}') \partial_k' G_{im}(\mathbf{x}, \mathbf{x}') \right) - G_{ij}(\mathbf{x}, \mathbf{x}') \partial_l' \left(c_{ljmk}(\mathbf{x}') \partial_m' u_k(\mathbf{x}') \right) \tag{11.15}$$
$$= -u_j(\mathbf{x}') \delta_{ji} \delta(\mathbf{x}' - \mathbf{x}) + G_{ij}(\mathbf{x}, \mathbf{x}') f_j(\mathbf{x}') \ .$$

It is obtained by using (11.13) and (11.14). By putting (11.15) into transformed (11.16), the sought identity is obtained.

According to (11.14), the function $u_i(\mathbf{x}) = G_{ki}(\mathbf{x}'', \mathbf{x})$ solves (11.13) for $f_i(\mathbf{x}) = \delta_{ik}\delta(\mathbf{x} - \mathbf{x}'')$. After substituting u_i and f_i in Eqs. (11.17), (11.18) and (11.19), and taking into account the boundary conditions for $G_{ki}(\mathbf{x}'', \mathbf{x})$, one gets

$$G_{ki}(\mathbf{x}'', \mathbf{x}) = G_{ik}(\mathbf{x}, \mathbf{x}'') .$$

If the stiffness tensor field is constant ($\mathbf{c}(\mathbf{x}) = \mathbf{c}^0$) on an infinite sample then the solution should be invariant under translations, i.e., for an arbitrary $\mathbf{v} \in V^\infty$ there occurs $\mathbf{G}(\mathbf{x} + \mathbf{v}, \mathbf{x}' + \mathbf{v}) = \mathbf{G}(\mathbf{x}, \mathbf{x}')$. In this case, the Green's tensor can be written as a function of only one argument $\mathbf{r} := \mathbf{x} - \mathbf{x}'$, i.e., $\mathbf{G} = \mathbf{G}(\mathbf{r})$.

For the infinite case, a more specific formula for $\mathbf{G}$ can be obtained. Let $F(f)$ and $F^{-1}(f)$ denote the Fourier and inverse Fourier transforms of a function $f : V^\infty \to$ R, respectively. By applying the operator F to both sides of $c^0_{ijkl}\partial_j\partial_k G_{lm}(\mathbf{r}) = -\delta_{im}\delta(\mathbf{r})$ and using

$$F(\partial_l f)(\mathbf{k}) = ik_l F(f)(\mathbf{k}) , \tag{11.20}$$

one gets

$$c^0_{ijkl}k_j k_k F(G_{lm})(\mathbf{k}) = \delta_{im}/(8\pi^3) .$$

Let $\mathbf{D}(\mathbf{k})$ denote the Christoffel matrix $D_{ik}(\mathbf{k}) := c^0_{ijkl}k_j k_k$. The Green's function can be written as

$$\mathbf{G}(\mathbf{r}) = \frac{1}{8\pi^3}F^{-1}(\mathbf{D}^{-1})(\mathbf{r}) = \frac{1}{8\pi^3}\int_{V_k^\infty} d_3\mathbf{k}\,(\mathbf{D}(\mathbf{k}))^{-1}\exp(i\mathbf{kr}) . \tag{11.21}$$

Elements of $\mathbf{D}$ are homogeneous polynomials of k_i of degree 2. Thus, the determinant of $\mathbf{D}$ and elements of the cofactor matrix are homogeneous polynomials of k_i of degree 6 and 4, respectively. Therefore, with the unit vector $\mathbf{e}$ in direction of $\mathbf{k}$ such that $\mathbf{k} = k\mathbf{e}$,

$$(\mathbf{D}(\mathbf{k}))^{-1} = (\mathbf{D}(\mathbf{e}))^{-1}/k^2 , \tag{11.22}$$

and the previous formula takes the form

$$\mathbf{G}(\mathbf{r}) = \frac{1}{8\pi^3}\int_{S_e^2+} d_2\mathbf{e}\,(\mathbf{D}(\mathbf{e}))^{-1}\int_{-\infty}^{+\infty} dk\,\exp(ik\mathbf{er})$$

with integration over the hemisphere S_e^2+ and the range of k from $-\infty$ to $+\infty$. Moreover, $\int_{-\infty}^{+\infty}\exp(ik\mathbf{er})dk = 2\pi\delta(i\mathbf{er}) = 2\pi\delta(i\mathbf{en})/r$, where $\mathbf{n}$ is a unit vector with the sense of $\mathbf{r}$, i.e., $\mathbf{r} = r\mathbf{n}$. Hence,

$$\mathbf{G}(\mathbf{r}) = \frac{1}{4\pi^2 r}\int_{S_e^2+} d_2\mathbf{e}\,(\mathbf{D}(\mathbf{e}))^{-1}\delta(i\mathbf{en}) .$$

Points of the hemisphere contributing to the integral are of the form $\mathbf{e} = \mathbf{v}^1\cos\phi + \mathbf{v}^2\sin\phi$, where $\mathbf{v}^1, \mathbf{v}^2$ are mutually orthogonal unit vectors, both perpendicular to $\mathbf{n}$. Therefore,

$$\mathbf{G}(\mathbf{r}) = \frac{1}{4\pi^2 r}\int_{-\pi/2}^{+\pi/2} d\phi\,(\mathbf{D}(\mathbf{v}^1\cos\phi + \mathbf{v}^2\sin\phi))^{-1} . \tag{11.23}$$

It is easy to notice that, for the infinite homogeneous medium, $\mathbf{G}$ is a symmetric tensor and an even function of $\mathbf{r}$ (i.e., $G_{ij}(\mathbf{r}) = G_{ji}(\mathbf{r}) = G_{ij}(-\mathbf{r})$) of the form $r^{-1}\times$(direction dependent factor).

We can proceed further only in special cases; e.g., the Green's tensor can be determined for an isotropic material with the stiffness tensor given by $\mathbf{c} = \mathbf{iso}(\kappa, \gamma)$. The tensor $\mathbf{c}$ can be expressed as $c_{ijkl} = \gamma I_{ijkl} + (\kappa - \gamma)\lambda\delta_{ij}\delta_{kl}/3$. Thus,

$$D_{ij}(\mathbf{e}) = \frac{\gamma}{2}\delta_{ij} + \frac{2\kappa + \gamma}{6}e_i e_j \ ,$$

and hence

$$D_{ij}^{-1}(\mathbf{e}) = \frac{2}{\gamma}(\delta_{ij} - \xi e_i e_j) \ , \tag{11.24}$$

where

$$\xi = \frac{2\kappa + \gamma}{2(\kappa + 2\gamma)} \ .$$

From (11.23),

$$\begin{aligned}\mathbf{G}_{ij}(\mathbf{r}) &= \frac{1}{2\pi^2\gamma r}\int_{-\pi/2}^{+\pi/2}[\delta_{ij} - \xi(v_i^1\cos\phi + v_i^2\sin\phi)(v_j^1\cos\phi + v_j^2\sin\phi)]d\phi \\ &= (1/4\pi\gamma r)\left(2\delta_{ij} - \xi(v_i^1 v_i^1 + v_i^2 v_j^2)\right) \ .\end{aligned}$$

Moreover, due to the orthogonality of $\mathbf{v}^1$, $\mathbf{v}^2$ and $\mathbf{n}$, there occurs $v_i^1 v_i^1 + v_i^2 v_j^2 + n_i n_j = \delta_{ij}$; hence,

$$G_{ij}(\mathbf{r}) = \frac{1}{4\pi\gamma\, r}\left((2 - \xi)\delta_{ij} + \xi n_i n_j\right)$$

which is the explicit expression for the Green's function in the isotropic case (Thompson, 1846).

As for the anisotropic cases, the analytic formula for the Green's tensor is known for hexagonal symmetry (Lifshitz, Rozentsveig, 1947, Kröner, 1953). For other symmetries, the calculations are limited by the problem of solving algebraic sextic equation (Head, 1979a,b); generally, the only way to determine the tensor is by numerical calculations (Barnett, 1972, see also Lie, Koehler, 1968, Mura, Kinoshita, 1971). ⊠

Let V be a representative area in a statistically homogeneous sample. Moreover, let $\mathbf{c}'$ denote the deviation of $\mathbf{c}$ from a constant field $\mathbf{c}^0$, i.e., $\mathbf{c}'(\mathbf{x}) := \mathbf{c}(\mathbf{x}) - \mathbf{c}^0$. The equilibrium condition $\partial_j\left(c_{ijkl}(\mathbf{x})\partial_k u_l(\mathbf{x})\right) = 0$ can be written in the form

$$c_{ijkl}^0\partial_j\partial_k u_l(\mathbf{x}) = -\partial_j\left(c_{ijkl}'(\mathbf{x})\partial_k u_l(\mathbf{x})\right) \ . \tag{11.25}$$

Now, the crucial step is to consider the right-hand side of the equation as a body force in the homogeneous material with elastic properties described by $\mathbf{c}^0$ (Lifshitz, Rozentsveig, 1946, Eimer, 1967, Kröner, 1967, Bolotin, Moskalenko, 1968b). The solution of Eq.(11.25) is given by the formula (11.16)

$$\begin{aligned}u_i(\mathbf{x}) &= \int_V G_{ij}(\mathbf{x}, \mathbf{x}')\partial_l'\left(c_{jlmk}'(\mathbf{x}')\partial_m' u_k(\mathbf{x}')\right)d_3\mathbf{x}' \\ &\quad - \int_{\partial V}[u_j(\mathbf{x}')c_{ljkm}^0\partial_k'G_{im}(\mathbf{x}, \mathbf{x}') - G_{ij}(\mathbf{x}, \mathbf{x}')c_{ljmk}^0\partial_m' u_k(\mathbf{x}')]n_l d_2\mathbf{x}' \ ,\end{aligned}$$

where the Green's function $\mathbf{G}$ corresponds to the tensor $\mathbf{c}^0$. After integrating the first component, and then using Gauss theorem, one gets the displacement

$$\begin{aligned}u_i(\mathbf{x}) &= -\int_V\left(\partial_l'G_{ij}(\mathbf{x}, \mathbf{x}')\right)c_{jlmk}'(\mathbf{x}')\partial_m' u_k(\mathbf{x}')d_3\mathbf{x}' \\ &\quad - \int_{\partial V}[u_j(\mathbf{x}')c_{ljkm}^0\partial_k'G_{im}(\mathbf{x}, \mathbf{x}') - G_{ij}(\mathbf{x}, \mathbf{x}')c_{ljmk}(\mathbf{x}')\partial_m' u_k(\mathbf{x}')]n_l d_2\mathbf{x}'\end{aligned}$$

and then the expression for the local strain

$$\varepsilon_{in}(\mathbf{x}) = -\int_V \left(\partial'_l\partial_{(n}G_{i)j}(\mathbf{x},\mathbf{x}')\right) c'_{jlmk}(\mathbf{x}')\varepsilon_{mk}(\mathbf{x}')d_3\mathbf{x}'$$
$$- \int_{\partial V}[u_j(\mathbf{x}')c^0_{ljkm}\partial'_k\partial_{(n}G_{i)m}(\mathbf{x},\mathbf{x}') - \partial_{(n}G_{i)j}(\mathbf{x},\mathbf{x}')c_{jlmk}(\mathbf{x}')\varepsilon_{mk}(\mathbf{x}')]n_l d_2\mathbf{x}' \ .$$
(11.26)

The last term of (11.26)

$$\varepsilon^0_{in}(\mathbf{x})$$
$$= -\int_{\partial V}[u_j(\mathbf{x}')c^0_{ljkm}\partial'_k\partial_{(n}G_{i)m}(\mathbf{x},\mathbf{x}') - \partial_{(n}G_{i)j}(\mathbf{x},\mathbf{x}')c_{jlmk}(\mathbf{x}')\varepsilon_{mk}(\mathbf{x}')]n_l d_2\mathbf{x}'$$

is of deterministic character.

Explanation
Assuming boundary conditions of the first kind, one has $G_{ij}(\mathbf{x},\mathbf{x}') = 0$ for $\mathbf{x}'$ in ∂V, and hence $\partial_k G_{ij}(\mathbf{x},\mathbf{x}') = 0$ on the boundary ∂V. Therefore the second component of the term ε^0 vanishes, and the first one is determined by the (deterministic) boundary condition. Similarly, for the boundary conditions of second kind, the second component of Eq.(11.26) vanishes because of $c^0_{ljkm}\partial'_k G_{im}(\mathbf{x},\mathbf{x}') = 0$ for $\mathbf{x}'$ on the boundary ∂V, whereas the third one is given by $c_{ljmk}(\mathbf{x}')\varepsilon_{mk}(\mathbf{x}')n_l(\mathbf{x}')_{|\partial V} = \sigma(\mathbf{x}')n_l(\mathbf{x}')_{|\partial V}$. ⊠

Thus, the equation (11.26) can be written as

$$\varepsilon(\mathbf{x}) = \varepsilon^0(\mathbf{x}) + (\mathsf{G}c'\varepsilon)(\mathbf{x}) \ ,$$
(11.27)

where the operator G is defined by

$$(\mathsf{G}\zeta)_{ij}(\mathbf{x}) = -\int_V \left(\partial'_l\partial_{(j}G_{i)k}(\mathbf{x},\mathbf{x}')\right)\zeta_{kl}(\mathbf{x}')\,d_3\mathbf{x}' \ , \quad \zeta \text{ in } \mathsf{T}^2.$$
(11.28)

The tensor ε^0 can be eliminated by ensemble averaging (11.27) and subtracting the result from the initial formula; with the argument $\mathbf{x}$ omitted,

$$<\varepsilon> = \varepsilon^0 + \mathsf{G}<c'\varepsilon>,$$
(11.29)

and

$$\varepsilon = <\varepsilon> + \mathsf{G}(c'\varepsilon - <c'\varepsilon>) \ .$$
(11.30)

The solution of the above equation with respect to ε can be obtained by consecutive substitution of successive approximations of ε (starting from $<\varepsilon>$) into the right-hand side of the equation (11.30). Let's define

$$\varepsilon_0 := <\varepsilon> \quad \text{and} \quad \varepsilon_n := \Gamma\varepsilon_{n-1} = \Gamma^n <\varepsilon> \ , \ n = 1, 2, \dots \ ,$$
(11.31)

where the non-local linear operator Γ acts on ζ (belonging to T^2) by

$$\Gamma\zeta = \mathsf{G}(c'\zeta - <c'\zeta>) \ .$$

There occurs $<\varepsilon_n> = 0$ for $n = 1, 2, \dots$. It can be easily verified that

$$\varepsilon = \sum_{n=0}^{\infty}\varepsilon_n = \sum_{n=0}^{\infty}\Gamma^n <\varepsilon> \ .$$

i.e., in general, the field ε is expressed in the form $\varepsilon = \mathsf{A} <\varepsilon>$ with operator A given by $\mathsf{A} = \sum_{n=0}^{\infty}\Gamma^n$.

11.2.2 General formula for effective tensor

Substituting the solution (11.2.1) for ε into the formula which defines $\mathbf{c}^*$

$$\mathbf{c}^* <\varepsilon> = <\mathbf{c}\varepsilon> = \mathbf{c}^0 , <\varepsilon> + <\mathbf{c}'\varepsilon> \tag{11.32}$$

with $<\cdot>$ denoting the ensemble average, one finally gets the expression for the effective stiffness tensor

$$\mathbf{c}^* = <\mathbf{c}A> = \mathbf{c}^0 + <\mathbf{c}' \sum_{n=0}^{\infty} \varGamma^n> . \tag{11.33}$$

In the more explicit form (up to the fourth correction),

$$
\begin{aligned}
\mathbf{c}^* = \mathbf{c}^0 \\
{}_{(1)} + <\mathbf{c}'> \\
{}_{(2)} + <\mathbf{c}'G\mathbf{c}'> - <\mathbf{c}'> G <\mathbf{c}'> \\
{}_{(3)} + <\mathbf{c}'G\mathbf{c}'G\mathbf{c}'> - <\mathbf{c}'G\mathbf{c}'> G <\mathbf{c}'> \\
\quad - <\mathbf{c}'> G <\mathbf{c}'G\mathbf{c}'> + <\mathbf{c}'> G <\mathbf{c}'> G <\mathbf{c}'> \\
{}_{(4)} + <\mathbf{c}'G\mathbf{c}'G\mathbf{c}'G\mathbf{c}'> - <\mathbf{c}'G\mathbf{c}'G\mathbf{c}'> G <\mathbf{c}'> - <\mathbf{c}'G\mathbf{c}'> G <\mathbf{c}'G\mathbf{c}'> \\
\quad + <\mathbf{c}'G\mathbf{c}'> G <\mathbf{c}'> G <\mathbf{c}'> - <\mathbf{c}'> G <\mathbf{c}'G\mathbf{c}'G\mathbf{c}'> \\
\quad + <\mathbf{c}'> G <\mathbf{c}'G\mathbf{c}'> G <\mathbf{c}'> + <\mathbf{c}'> G <\mathbf{c}'> G <\mathbf{c}'G\mathbf{c}'> \\
\quad - <\mathbf{c}'> G <\mathbf{c}'> G <\mathbf{c}'> G <\mathbf{c}'> + ...
\end{aligned}
\tag{11.34}
$$

(cf. Bolotin and Moskalenko, 1968a,b, or Beran and McCoy, 1970).

The result (11.33) constitutes a formal solution of the considered problem. Its relatively complicated character should be noticed. The "effective" stress–strain relation is non–local, i.e., $<\sigma>(\mathbf{x})$ is related to $<\varepsilon>(\mathbf{x})$ via $\int \mathbf{c}^*(\mathbf{x}, \mathbf{x}') <\varepsilon>(\mathbf{x}') d_3\mathbf{x}'$. Generally, a polycrystal will not be macroscopically homogeneous. With additional assumption of macro–homogeneity, the operator $\mathbf{c}^*$ depends on only one spatial variable $\mathbf{r} = \mathbf{x}' - \mathbf{x}$, i.e., $\mathbf{c}^* = \mathbf{c}^*(\mathbf{r})$. It can be seen from (11.34) that in order to get strict results, the ('correlation') functions of the type $<\mathbf{c}'G\mathbf{c}'G...\mathbf{c}'G\mathbf{c}'>$ of all orders would have to be known. In practice, statistical information about the sample's structure is not complete, i.e., only the functions of the lowest order are available. Therefore, the series (11.34) has to be truncated after a few first components or simplifying assumptions are necessary. One may ask whether the series (11.33) is convergent. As will be seen below, in simple cases the series is geometric and its convergence is obvious.

Dual form of effective tensor

The solution (11.34) can be written in the dual form, i.e., based on the compliance tensors $\mathbf{s}$ and $\mathbf{s}^*$. By defining σ^0 as $\sigma^0 := \mathbf{c}^0\varepsilon^0$, using Hooke's law $\varepsilon = \mathbf{s}\sigma$ and multiplying both sides of (11.27) by $\mathbf{c}^0$, one gets

$$\mathbf{c}^0 \mathbf{s}\sigma = \sigma^0 + \mathbf{c}^0 G\mathbf{c}'\mathbf{s}\sigma . \tag{11.35}$$

Let $\mathbf{s}' := \mathbf{s} - \mathbf{s}^0$, where $\mathbf{s}^0 := (\mathbf{c}^0)^{-1}$. Substituting $\mathbf{c}^0\mathbf{s}$ and $\mathbf{c}'\mathbf{s}$ by

$$\mathbf{c}^0 \mathbf{s} = \mathbf{c}^0 (\mathbf{s}^0 + \mathbf{s}') = \mathbf{I} + \mathbf{c}^0 \mathbf{s}'$$

and
$$\mathbf{c's} = (\mathbf{c} - \mathbf{c}^0)\mathbf{s} = \mathbf{I} - \mathbf{c}^0\mathbf{s} = \mathbf{c}^0\mathbf{s}^0 - \mathbf{c}^0\mathbf{s} = -\mathbf{c}^0\mathbf{s'} ,$$
the equation (11.35) can be transformed to the form analogous to (11.27), i.e.

$$\sigma = \sigma^0 + \mathsf{H}\mathbf{s'}\sigma , \tag{11.36}$$

where H is an operator defined by

$$\mathsf{H} := -\mathbf{c}^0 - \mathbf{c}^0 \mathsf{G} \mathbf{c}^0 . \tag{11.37}$$

Proceeding like in the case of the solution (11.34), one obtains analogous result with $\mathbf{c}^*$, $\mathbf{c}^0$, $\mathbf{c'}$ and G replaced respectively by $\mathbf{s}^*$, $\mathbf{s}^0$, $\mathbf{s'}$ and H

$$\mathbf{s}^* = \mathbf{s}^0 + <\mathbf{s'}> + <\mathbf{s'}\mathsf{H}\mathbf{s'}> - <\mathbf{s'}> \mathsf{H} <\mathbf{s'}> +... \tag{11.38}$$

(Dederichs and Zeller, 1972). ⊠

Based on (11.34), it is clear that the first order approximation of $\mathbf{c}^*$ is equivalent to the Voigt average $\mathbf{c}^* = <\mathbf{c}>$. With Eq.(11.38), the same approximation for $\mathbf{s}^*$ gives the Reuss average $\mathbf{s}^* = <\mathbf{s}>$. It is worth noticing that these first corrections are influenced by orientation distributions, whereas grain shapes (and all types of correlations) are immaterial.

11.2.3 The comparison field $\mathbf{c}^0$ and self–consistent approach

It should be stressed once more that the tensor $\mathbf{c}^0$, and subsequently, the related operator G, have not been specified. Hence, there is some arbitrariness of choice. Assuming that $\mathbf{c}^0 = <\mathbf{c}>$ (i.e., $<\mathbf{c'}> = <\mathbf{c}> - \mathbf{c}^0 = 0$), the formula (11.34) for $\mathbf{c}^*$ becomes considerably simpler

$$\begin{aligned} \mathbf{c}^* = <\mathbf{c}> &+ <\mathbf{c'}\mathsf{G}\mathbf{c'}> + <\mathbf{c'}\mathsf{G}\mathbf{c'}\mathsf{G}\mathbf{c'}> \\ &+ <\mathbf{c'}\mathsf{G}\mathbf{c'}\mathsf{G}\mathbf{c'}\mathsf{G}\mathbf{c'}> - <\mathbf{c'}\mathsf{G}\mathbf{c'}> \mathsf{G} <\mathbf{c'}\mathsf{G}\mathbf{c'}> +... \end{aligned} \tag{11.39}$$

That is a tempting substitution. The result, however, is not strict but only a useful approximation because there is no formal justification of such choice of $\mathbf{c}^0$.

The tensor $\mathbf{c}^0$ is determined numerically by applying iteration (self–consistent) scheme with the comparison field $\mathbf{c}^0_{n+1}$ in $n+1$ step equal to effective tensor $\mathbf{c}^*_n$ obtained from n-th step. Formally, the method leads to an equation for $\mathbf{c}^*$. Let the relationship (11.33) between $\mathbf{c}^*$ and the chosen $\mathbf{c}^0$ be written briefly as $\mathbf{c}^* = \mathfrak{F}(\mathbf{c}^0)$. The iteration procedure is reduced to consecutive substitutions $\mathbf{c}^0_{n+1} = \mathbf{c}^*_n = \mathfrak{F}(\mathbf{c}^0_n)$ with the proper effective tensor $\mathbf{c}^*$ being the limit of the series of $\mathbf{c}^*_n$ determined that way, (i.e., $\mathbf{c}^* = \lim_{n\to\infty} \mathbf{c}^*_n$). In other words, $\mathbf{c}^*$ should be a fixed point of the mapping $\mathfrak{F}$, i.e., it should satisfy the condition

$$\mathbf{c}^* = \mathfrak{F}(\mathbf{c}^*) .$$

This is the equation of self–consistency. In general form it is complicated; however, various approximations are possible. One of them based on a simplification of the action of the operator G will be considered below.

11.2.4 Properties of the operator G

From the viewpoint of performing calculations, the simplest approach is to approximate G by the operator corresponding to the infinite medium. This approximation is assumed in further considerations.

The operator G is defined on the basis of the Green's function by (11.28). The derivatives of **G** which appear in the definition are locally integrable functions and, therefore, there exists the distribution $\mathcal{G}$ such that $(G\zeta)(\mathbf{x}) = \int \mathcal{G}(\mathbf{r})\zeta(\mathbf{x}+\mathbf{r})d_3\mathbf{r}$. The distribution can be split into $\mathcal{G}(\mathbf{r}) = \mathcal{G}^0\delta(\mathbf{r}) + \mathcal{G}^1(\mathbf{r})$ with non–zero constant tensor $\mathcal{G}^0$ (belonging to T^4). For a unit vector **n**, the dependence of $\mathcal{G}^1(r\mathbf{n})$ on the distance r has the form $r^{-3}\times$(coefficient depending on **n**), i.e.,

$$\mathcal{G}^1(r\mathbf{n}) = r^{-3}\mathcal{G}^1(\mathbf{n}) \ . \tag{11.40}$$

Based on this property, it becomes clear that the effective quantity depends on geometry (microstructure) only up to the scale. The transformation $\mathbf{r} \rightarrow \alpha\mathbf{r}$ does not change the final result since the integrals in (11.34) containing $\mathcal{G}(\mathbf{r})$ are not influenced by such transformation. This is obvious for terms involving $\mathcal{G}^0\delta(\mathbf{r})$. Also the product $\mathcal{G}^1(\mathbf{r})d_3\mathbf{r}$ is invariant because $\mathcal{G}^1(\alpha\mathbf{r})d_3(\alpha\mathbf{r}) = \alpha^{-3}\mathcal{G}^1(\mathbf{r})\alpha^3 d_3\mathbf{r} = \mathcal{G}^1(\mathbf{r})d_3\mathbf{r}$.

This scale-independence is related to the error caused by the mentioned approximation of actual G by operator for infinite medium. It is clear that the error does not depend on the absolute size of the sample but it is rather connected with the ratio between the characteristic size of inhomogeneities and dimensions of the sample.

The integral of the 'tail' $\mathcal{G}^1(\mathbf{r})$ of the distribution $\mathcal{G}$ over the unit sphere S_r^2 satisfies $\int_{S_r^2} \mathcal{G}^1(\mathbf{r})d_2\mathbf{r} = 0$. Because of that, the terms of the series (11.33) with correlation functions which do not depend on spatial directions are eliminated. If none of the correlation functions depend on directions and the orientation distribution is uniform ($f = 1$), the material is described as *strongly isotropic* (Bolotin and Moskalenko, 1969). In particular, the so–called *perfectly disordered material* with point-like grains, uniform orientation distribution and no spatial correlations considered by Kröner (1967) is strongly isotropic.

⊡ *Derivation and properties of $\mathcal{G}^0$ and $\mathcal{G}^1$*
    The goal of this fragment is to derive the expression for $\mathcal{G}^0$ in the isotropic case and to justify statements given above concerning the operator G. In general case, to obtain the derivatives of Green's function, numerical calculations are necessary (p. Mura, Kinoshita, 1971, Barnett, 1972). Some useful properties, however, can be obtained analytically. The operator G is defined on the basis of Green's function by Eq.(11.28). Taking into account that $\partial'_l\partial_{(j}G_{i)k}(\mathbf{x}' - \mathbf{x}) = -\partial_l\partial_{(j}G_{i)k}(\mathbf{x}' - \mathbf{x})$, the defining formula can be written as

$$(G\zeta)_{ij}(\mathbf{x}) := \int_V [\partial_l\partial_{(j}G_{i)}(\mathbf{x}' - \mathbf{x})]\zeta_{kl}(\mathbf{x}')d_3\mathbf{x}' \ , \quad \zeta \in \mathsf{T}^2$$

The tensor $\mathcal{G}^0$ can be obtained by calculating the integral for a ball $K(d)$ with the center at $\mathbf{r} = \mathbf{0}$ and of the radius d; with d approaching zero there occures $\lim_{d\to 0} \int_{K(d)} [\partial_l\partial_{(j}G_{i)k}(\mathbf{r})]\zeta_{kl}(\mathbf{x}+\mathbf{r})d_3\mathbf{r} = \mathcal{G}^0_{ijkl}\zeta_{kl}(\mathbf{x})$. Hence,

$$\mathcal{G}^0_{ijkl} = \lim_{d\to 0} \int_{K(d)} \partial_l\partial_{(j}G_{i)k}(\mathbf{r})d_3\mathbf{r} \ .$$

Let us calculate the integral $\mathbf{E}(d)$ given by $E_{ijkl}(d) := \int_{K(d)} \partial_l \partial_{(j} G_{i)k}(\mathbf{x}) d_3\mathbf{x}$. Taking consecutively Fourier transform and inverse Fourier transform of the integrand gives

$$E_{ijkl}(d) = \int_{K(d)} F^{-1}\left(F(\partial_l \partial_{(j} G_{i)k})\right) d_3\mathbf{x} \ .$$

Due to (11.20) and (11.21), $\mathbf{E}$ can be expressed through the inverse of the Christoffel matrix $\mathbf{D}$

$$E_{ijkl}(d) = -\int_{K(d)} F^{-1}\left(k_l k_{(j} F(G_{i)k})\right) d_3\mathbf{x} = -\frac{1}{8\pi^3} \int_{K(d)} F^{-1}\left(k_l k_{(j} D_{i)k}^{-1}(\mathbf{k})\right) d_3\mathbf{x} \ .$$

The explicit form of this formula is

$$E_{ijkl}(d) = -\frac{1}{8\pi^3} \int_{K(d)} d_3\mathbf{x} \int_{V_k^\infty} k_l k_{(j} D_{i)k}^{-1}(\mathbf{k}) \exp\left(i\mathbf{kx}\right) d_3\mathbf{k} \ .$$

Because of (11.22), $k_l k_{(j} D_{i)k}^{-1}(\mathbf{k}) = e_l e_{(j} D_{i)k}^{-1}(\mathbf{e}) =: B_{ijkl}(\mathbf{e})$. The expression for $\mathbf{E}(d)$ takes the form

$$\mathbf{E}(d) = -\frac{1}{8\pi^3} \int_{K(d)} d_3\mathbf{x} \int_{S_e^2} d_2\mathbf{e} \int_0^\infty \mathbf{B}(\mathbf{e}) k^2 \exp\left(i k \mathbf{ex}\right) dk \ .$$

Now, there occurs $\int_{-\infty}^\infty k^2 \exp\left(iky\right) dk = -2\pi\delta^{(2)}(y)$, where $\delta^{(2)}$ is the second derivative of the δ distribution with respect to its argument. Moreover, one has $\int_{K(d)} \delta^{(2)}(\mathbf{ex}) d_3\mathbf{x} = -2\pi$. This leads to

$$\mathbf{E}(d) = -\frac{1}{4\pi} \int_{S_e^2} \mathbf{B}(\mathbf{e}) d_2\mathbf{e} \ .$$

It is visible now that $\mathbf{E}$ is actually independent of d. Thus, the singular part $\mathcal{G}^0$ of the distribution $\mathcal{G}$, is given by

$$\mathcal{G}^0 = -\frac{1}{4\pi} \int_{S_e^2} \mathbf{B}(\mathbf{e}) d_2\mathbf{e} \ .$$

For isotropic medium $\mathbf{c} = \mathbf{iso}(\kappa, \gamma)$. According to Eq.(11.24), $D_{ij}^{-1}(\mathbf{e}) = 2(\delta_{ij} - \xi e_i e_j)/\gamma$ and thus, the tensor $\mathbf{B}$ is given by $B_{ijkl}(\mathbf{e}) = 2(e_l e_{(j} \delta_{i)k} - \xi e_i e_j e_k e_l)/\gamma$. Hence,

$$\mathcal{G}_{ijkl}^0 = (1/2\pi\gamma) \int_{S_e^2} (\xi e_i e_j e_k e_l - e_l e_{(j} \delta_{i)k}) d_2\mathbf{e} \ .$$

The integrals of $e_i e_j e_k e_l$ and $e_i e_j$ with respect to all directions are isotropic and the invariant $e_i e_i$ is equal to 1. Therefore,

$$\int_{S_e^2} e_i e_j e_k e_l d_2\mathbf{e} = \frac{4\pi}{15}(\delta_{ij}\delta_{kl} + \delta_{ik}\delta_{jl} + \delta_{il}\delta_{jk}) \quad \text{and} \quad \int_{S_e^2} e_i e_j d_2\mathbf{e} = \frac{4\pi}{3}\delta_{ij} \ .$$

Hence,

$$\mathcal{G}_{ijkl}^0 = \frac{2}{15\gamma} (\xi \delta_{ij}\delta_{kl} + (2\xi - 5)I_{ijkl})$$

or

$$\mathcal{G}^0 = \frac{2}{15\gamma}\mathbf{iso}\left(5(\xi - 1), 2\xi - 5\right) \ . \tag{11.41}$$

The last expression will be used in our example calculations.

As for $\mathcal{G}^1(\mathbf{r})$, it can be shown that $\int_{S_r^2} \mathcal{G}^1(\mathbf{r}) d_2 \mathbf{r} = 0$. Analogously to the calculation of $\mathbf{E}$, one has

$$\int_{S_r^2} \mathcal{G}^1(\mathbf{r}) d_2\mathbf{r} = -\frac{1}{8\pi^3} \int_{S_r^2} d_2\mathbf{r} \int_{V_k^\infty} k_l k_{(j} D_{i)k}^{-1}(\mathbf{k}) \exp(i\mathbf{kr}) \, d_3\mathbf{k} =$$

$$-\frac{1}{8\pi^3} \int_{S_r^2} d_2\mathbf{r} \int_{S_e^2} d_2\mathbf{e} \int_0^\infty \mathbf{B}(\mathbf{e}) k^2 \exp(ikren) \, dk = \frac{1}{8\pi^3} \int_{S_e^2} \mathbf{B}(\mathbf{e}) \partial_j \partial_j L(\mathbf{e}, r) \, d_2\mathbf{e}$$

where $L(\mathbf{e}, r)$ is given by

$$L(\mathbf{e}, r) := \int_{S_r^2} d_2\mathbf{r} \int_0^\infty \exp(ikren) \, dk \ .$$

Since $\int_0^\infty \exp(ikx) dk = \pi\delta(x)$ and $\delta(cx) = \delta(x)/|c|$,

$$L(\mathbf{e}, r) = \frac{\pi}{r} \int_{S_r^2} d_2\mathbf{r} \, \delta(\mathbf{en}) \ .$$

For non–zero r, one has $\partial_j \partial_j (1/r) = 0$; hence, $\partial_j \partial_j L(\mathbf{e}, r) = 0$ and thus, there occurs $\int_{S_r^2} \mathcal{G}^1(\mathbf{r}) d_2\mathbf{r} = 0$.

It is clear from (11.23) that $\mathbf{G}$ is of the form $r^{-1} \times$(direction dependent factor). Therefore, the derivatives $\partial_l \partial_j G_{ik}(\mathbf{r})$ and the term $\mathcal{G}^1$ have the form $r^{-3} \times$(factor depending on the direction of vector $\mathbf{r}$ but not on r). $\boxtimes$

Example direct summation

In some very simple cases, it is possible to proceed with direct summation of the series (11.34) and (11.38). Let us give a constructive example of such procedure for the series (11.39). It is a bit cumbersome but has educational value. For the two–point density function independent of the direction of the vector $\mathbf{r}$, the integral containing $\mathcal{G}^1$ vanishes. With tensor $\mathbf{c}$ at a given point determined by the orientation of the crystallite containing that point, the explicit form of the second term of (11.39) is given by $<\mathbf{c}'\mathbf{Gc}'> = \mathcal{M}^f (\mathbf{c}'\mathcal{G}^0\mathbf{c}')$. Similar approach to higher order corrections gives

$$<\mathbf{c}' \underbrace{\mathbf{Gc}'...\mathbf{Gc}'}_{n \text{ factors}}> \mathbf{G} <\mathbf{c}' \underbrace{\mathbf{Gc}'...\mathbf{Gc}'}_{m \text{ factors}}> = \mathcal{M}^f(\mathbf{c}' \underbrace{\mathcal{G}^0\mathbf{c}'...\mathcal{G}^0\mathbf{c}'}_{n \text{ factors}})\mathcal{G}^0\mathcal{M}^f(\mathbf{c}' \underbrace{\mathcal{G}^0\mathbf{c}'...\mathcal{G}^0\mathbf{c}'}_{m \text{ factors}})$$

so the final solution is given by the series (11.34) with the operator $\mathbf{G}$ and $<\cdot>$ replaced by the tensor $\mathcal{G}^0$ and $\mathcal{M}^f$, respectively.

For uniform orientation distribution, cubic crystal symmetry with $\mathbf{c} = \mathbf{cub}(\kappa, \rho, \gamma)$, and isotropic $\mathbf{c}^0 = \mathbf{iso}(\kappa^0, \gamma^0)$, the above formula can be transformed further

$$\mathcal{M}(\underbrace{(\mathbf{c} - \mathbf{c}^0) \, \mathcal{G}^0(\mathbf{c} - \mathbf{c}^0)...\mathcal{G}^0(\mathbf{c} - \mathbf{c}^0))}_{n \text{ factors}}$$

$$= \mathbf{iso}\left((\kappa - \kappa^0)^{n+1} \Xi_k^n, \ \Xi^n (2(\rho - \gamma^0)^{n+1} + 3(\gamma - \gamma^0)^{n+1})/5 \right),$$

where

$$\Xi_k = -1/(\kappa^0 + 2\gamma^0) \quad \text{and} \quad \Xi := -\frac{2}{5} \frac{\kappa^0 + 3\gamma^0}{\gamma^0(\kappa^0 + 2\gamma^0)} \tag{11.42}$$

are parameters of the tensor $\mathcal{G}^0 = \mathbf{iso}(\Xi_k, \Xi)$ calculated from (11.41). For $\mathbf{c}^0$ chosen to be the Voigt average of $\mathbf{c} = \mathbf{cub}(\kappa, \rho, \gamma)$, one has $\kappa^0 = \kappa$, $\gamma^0 = \gamma^V = (2\rho + 3\gamma)/5$ or $\mathbf{c}^0 = \mathbf{iso}(\kappa, \gamma^V)$. Thus, there occurs

$$<\mathbf{c}' \underbrace{\mathbf{Gc}'\mathbf{G}...\mathbf{Gc}'}_{n \text{ factors}}> = \mathbf{iso}\left(0,\, 6X^n \left(3^n - (-2)^n\right)(\rho - \gamma)/25\right) , \tag{11.43}$$

where $X := \Xi(\rho - \gamma)/5$ and Ξ, in this particular case, takes the form

$$\Xi = -2\left((4\rho + 6\gamma + 5\kappa)^{-1} + (2\rho + 3\gamma)^{-1}\right) . \tag{11.44}$$

The second parameter γ_n of the n-th correction term to $\mathbf{c}^*$ in the series (11.39) is

$$\gamma_n = 6(\rho - \gamma)X^n/5 . \tag{11.45}$$

This can be proved by induction[4].

The corrections (11.45) constitute a geometric sequence and the series can be easily summed. For positive definite $\mathbf{c} = \mathbf{cub}(\kappa, \rho, \gamma)$ (i.e., $\kappa, \rho, \gamma > 0$) the ratio of the sequence is less than 1 and the series is convergent. The final result for the effective tensor is

$$\mathbf{c}^* = \mathbf{iso}\left(\kappa, \gamma^V + \frac{1}{5}\frac{6\Xi(\rho - \gamma)^2}{5 - \Xi(\rho - \gamma)}\right) \tag{11.47}$$

with Ξ given by (11.44).

Summation for $\mathbf{c}^0 = <\mathbf{s}>^{-1}$
Assumption that $\mathbf{c}^0 = <\mathbf{s}>^{-1}$ and a calculation similar to that used in derivation of Eq.(11.47) lead to

$$\mathbf{s}^* = \mathbf{iso}\left(\frac{1}{\kappa}, \frac{1}{\gamma^R} + \frac{1}{5}\frac{6\Theta_s\left(1/\rho - 1/\gamma\right)^2}{5 - \Theta_s\left(1/\rho - 1/\gamma\right)}\right) , \tag{11.48}$$

where $\gamma^R = (5\rho\gamma)/(2\gamma + 3\gamma)$, $\Theta_s = -\gamma^R - (\gamma^R)^2 \Xi_s$ and Ξ_s is second parameter of isotropic tensor $\mathcal{G}^0$ corresponding to $\mathbf{c}^0 = <\mathbf{s}>^{-1} = \mathbf{iso}(\kappa, \gamma^R)$, i.e.,

$$\Xi_s = 2\frac{2\gamma + 3\rho}{5}\left(\frac{1}{\kappa(2\gamma + 3\rho) + 10\rho\gamma} + \frac{1}{5\rho\gamma}\right) .$$

The result (11.48) differs from (11.47), i.e., the condition of reciprocity of $\mathbf{c}^*$ and $\mathbf{s}^*$ is not satisfied. $\boxtimes$

---

[4] For $n = 1$ and $n = 2$ the relation follows directly from (11.43). Let the second parameter of (11.43) be denoted by φ_n. The series (11.39) is is constructed in such a way that

$$\gamma_m = \varphi_m - \sum_{i=1}^{m-2} \varphi_i \Xi \gamma_{m-i-1} . \tag{11.46}$$

If formula (11.45) is valid for $n = 1, 2, ..., m - 2$, it is also valid for $n = m$. To verify this, we substitute φ_n and γ_n in the right-hand side of Eq.(11.46)

$$\gamma_m = (6/5)(\rho - \gamma)X^m\left((3^m - (-2)^m)/5 - \sum_{i=1}^{m-2} 6(3^i - (-2)^i)/5\right) .$$

Since $\sum_{i=1}^{m-2} 3^i = (3^m - 9)/6$ and $\sum_{i=1}^{m-2} 2^i = ((-2)^m - 4)/6$ we get $\gamma_m = (6/5)(\rho - \gamma)X^m$, i.e., the relation (11.45).

11.2.5 Singular approximation

In the above example it was possible to perform the calculations because the second term of $\mathcal{G}(\mathbf{r}) = \mathcal{G}^0 \delta(\mathbf{r}) + \mathcal{G}^1(\mathbf{r})$ was omitted. This is the basis of a more general method – the so called singular approximation in which the spatial part of $\mathcal{G}(\mathbf{r})$ is approximated by δ–distribution, i.e., $\mathcal{G}(\mathbf{r}) \approx \mathcal{G}^0 \delta(\mathbf{r})$ (Fokin, 1972). In such case, the equation (11.27) has the local form

$$\varepsilon = \varepsilon^0 + \mathcal{G}^0 \mathbf{c}' \varepsilon \ . \tag{11.49}$$

Therefore, the singular approximation is equivalent to discarding the information contained in many-point functions; only the orientation distribution is taken into account. From (11.49) follows

$$\varepsilon = (\mathbf{I} - \mathcal{G}^0 \mathbf{c}')^{-1} \varepsilon^0 \ .$$

After averaging $(<\varepsilon> = <(\mathbf{I} - \mathcal{G}^0 \mathbf{c}')^{-1}> \varepsilon^0)$ we get the expression for ε

$$\varepsilon = (\mathbf{I} - \mathcal{G}^0 \mathbf{c}')^{-1} <(\mathbf{I} - \mathcal{G}^0 \mathbf{c}')^{-1}>^{-1} <\varepsilon> \ .$$

Based on the definition $\mathbf{c}^* <\varepsilon> = <\mathbf{c}\varepsilon>$, the effective tensor is obtained

$$\mathbf{c}^* = <\mathbf{c}(\mathbf{I} - \mathcal{G}^0 \mathbf{c}')^{-1}> <(\mathbf{I} - \mathcal{G}^0 \mathbf{c}')^{-1}>^{-1} \tag{11.50}$$

(Fokin, 1972). Using (11.36) and (11.37), one can write down the dual result

$$\mathbf{s}^* = <\mathbf{s}(\mathbf{I} - \mathcal{H}^0 \mathbf{s}')^{-1}> <(\mathbf{I} - \mathcal{H}^0 \mathbf{s}')^{-1}>^{-1} \ , \tag{11.51}$$

where $\mathcal{H}^0 = -\mathbf{c}^0 - \mathbf{c}^0 \mathcal{G}^0 \mathbf{c}^0$. The tensors $\mathbf{c}^*$ and $\mathbf{s}^*$ calculated from Eqs. (11.50) and (11.51), respectively, are mutually reciprocal. It follows from the easy to prove relations

$$\mathbf{c}(\mathbf{I} - \mathcal{G}^0 \mathbf{c}')^{-1} = (\mathbf{I} - \mathcal{H}^0 \mathbf{s}')^{-1} \mathbf{c}^0 \quad \text{and} \quad \mathbf{s}(\mathbf{I} - \mathcal{H}^0 \mathbf{s}')^{-1} = (\mathbf{I} - \mathcal{G}^0 \mathbf{c}')^{-1} \mathbf{s}^0 \ .$$

Example:
As before the special case of cubic crystal symmetry and uniform orientation distribution will be considered. Relations (11.50) and (11.51) lead to

$$\mathbf{c}^* = \mathbf{iso} \left(\kappa, \frac{\chi + \gamma^V}{\chi + \gamma^R} \gamma^R \right) \ , \tag{11.52}$$

where $\gamma^V = (2\rho + 3\gamma)/5$, $\gamma^R = (5\rho\gamma)/(2\gamma + 3\gamma)$, $\chi := -\rho\gamma/(\gamma^0 + \varXi^{-1})$, with γ^0 and $\varXi$ being second parameters of isotropic (**iso**) tensors $\mathbf{c}^0$ and $\mathcal{G}^0$, respectively. Using Eq.(11.42) $(\varXi = -2(\kappa + 3\gamma^0)/(5\gamma^0(\kappa + 2\gamma^0)))$ the expression (11.52) can be transformed to

$$\mathbf{c}^* = \mathbf{iso} \left(\kappa, \frac{4\gamma^V(\gamma^0)^2 + 3(2\rho\gamma + \kappa\gamma^V)\gamma^0 + 2\kappa\rho\gamma}{4\gamma^R(\gamma^0)^2 + 3(2\rho\gamma + \kappa\gamma^R)\gamma^0 + 2\kappa\rho\gamma} \gamma^R \right) \ .$$

Now, it is easy to verify that
–for $\gamma^0 \to 0$ one gets Reuss' result,
–for $\gamma^0 \to \infty$ one gets Voigt's result,
–for $\gamma^0 = \gamma^V$ one gets the result (11.47),
–for $\gamma^0 = \gamma^R$ one gets the result (11.48).
Moreover,
–for $\gamma^0 = \rho$ one gets the first of the so–called Hashin–Shtrikman bounds,
–for $\gamma^0 = \gamma$ one gets the second of the Hashin–Shtrikman bounds (Hashin & Shtrikman, 1962, Hill, 1963).
Finally,
–for $\gamma^0 = \gamma^*$ one gets an algebraic equation for γ^*

$$4(\gamma^*)^3 + (3\kappa + 2\rho)(\gamma^*)^2 - (\kappa + 6\rho)\gamma\gamma^* - 2\kappa\rho\gamma = 0 \ . \tag{11.53}$$

The equation (11.53) (times an immaterial factor) was obtained by Hershey (1954) based on the model of spherical inhomogeneity surrounded by homogeneous matrix with effective properties. (See also Marutake, 1956 and Kröner, 1958). Due to the positive definiteness of the stiffness tensor $\mathbf{c}$ ($\kappa, \rho, \gamma > 0$), the equation (11.53) is casus irreducibilis and has three real roots. However, the solution γ^* is unique; that is because the sum of the roots $-(3\kappa+2\rho)/4$ is negative and their product is $\kappa\rho\gamma/2$ is positive, and hence only one of the roots of the equation is positive (what is required by positive definiteness of $\mathbf{c}^*$). The form of this root is

$$\gamma^* = \frac{1}{12}\left(-k + 2p\cos\left(\frac{1}{3}\arccos(q/p^3)\right)\right) \ ,$$

where $p := \left(\kappa^2 + 12\gamma(k + 6\rho)\right)^{1/2}$, $q := -k^3 - 18\gamma(3\kappa^2 - 4\kappa\rho + 12\rho^2)$ and $k := 3\kappa + 2\rho$ (see also Gairola and Kröner, 1981, and Morawiec, 1996b). ⊠

11.2.6 $\mathcal{T}$–matrix formalism

There is a possibility to get an insight into the problem from a different point of view by using the so–called $\mathcal{T}$–matrix approach adopted from the scattering theory (Zeller & Dederichs, 1973). The basic equations used above can be rewritten in a form involving the field of $\mathcal{T}$ which is defined by the relation

$$\mathcal{T} = \mathbf{c}' + \mathbf{c}'\mathsf{G}\mathcal{T} \ , \tag{11.54}$$

where $\mathcal{T}$ is a tensor operator with symmetries of $\overline{\mathsf{T}}^4$. Substitutions of (11.54) into itself starting from $\mathcal{T} = 0$ lead to

$$\mathcal{T} = \mathbf{c}' \sum_{i=0}^{\infty} (\mathsf{G}\mathbf{c}')^i \ . \tag{11.55}$$

Because of (11.54), the field

$$\varepsilon = (\mathbf{I} + \mathsf{G}\mathcal{T})\varepsilon^0 \tag{11.56}$$

satisfies equation (11.27). By taking the ensemble average of (11.56), ε^0 is eliminated; first,

$$<\varepsilon> = (\mathbf{I} + \mathsf{G} <\mathcal{T}>)\varepsilon^0 \tag{11.57}$$

and, formally, for non–singular $\mathbf{I} + \mathsf{G} <\mathcal{T}>$, one gets

$$\varepsilon = (\mathbf{I} + \mathsf{G}\mathcal{T})(\mathbf{I} + \mathsf{G} <\mathcal{T}>)^{-1} <\varepsilon> .$$

Putting this strain field ε into the expression defining $\mathbf{c}^*$ (i.e., into $\mathbf{c}^* <\varepsilon> = \mathbf{c}^0 <\varepsilon> + <\mathbf{c}'\varepsilon>$), the formula for the effective stiffness tensor expressed via $<\mathcal{T}>$ is obtained

$$\mathbf{c}^* = \mathbf{c}^0 + <\mathcal{T}> (\mathbf{I} + \mathsf{G} <\mathcal{T}>)^{-1} . \tag{11.58}$$

◻ *Example*
The calculations are simple when the above approach is applied to the strongly isotropic polycrystal with cubic crystal symmetry. Taking $\mathbf{c}^0 = <\mathbf{c}>$ and using (11.55) and (11.43),

$$<\mathcal{T}> = \mathbf{iso}\left(0, \frac{6\Xi(\rho-\gamma)^2}{(5 - 3\Xi(\rho-\gamma))(5 + 2\Xi(\rho-\gamma))}\right) .$$

Here, the action of the operator G on $<\mathcal{T}>$ is reduced to the multiplication of the second parameters of isotropic tensors. Taking this into account and using Eq.(11.58), the result given by (11.47) is obtained. ⊠

11.2.7 Self-consistent method based on $\mathcal{T}$–matrix

Application of the self–consistent procedure to Eq.(11.58) (by substitution $\mathbf{c}^0 = \mathbf{c}^*$) leads to $<\mathcal{T}> (\mathbf{I} + \mathsf{G} <\mathcal{T}>)^{-1} = 0$. Moreover, the relation (11.29) gives

$$\varepsilon^0 = <\varepsilon> -\mathsf{G}(<\mathbf{c}\varepsilon> -\mathbf{c}^* <\varepsilon>) = <\varepsilon> . \tag{11.59}$$

Subsequently, due to Eq.(11.57) and arbitrariness of $<\varepsilon>$, this gives the equation for the effective tensor

$$<\mathcal{T}> = 0 , \tag{11.60}$$

where (according to Eq.(11.55)) $\mathcal{T}$ has the form

$$\mathcal{T} = (\mathbf{c} - \mathbf{c}^*) \sum_{i=0}^{\infty} (\mathsf{G}(\mathbf{c} - \mathbf{c}^*))^i . \tag{11.61}$$

Again, the above equations can be solved with respect the effective tensor $\mathbf{c}^*$ in the special case of strongly isotropic medium. (See the example below.)

◻ *Example*
Let's consider again the strongly isotropic polycrystal with cubic crystal symmetry. Based on Eq.(11.61) and relations analogous to (11.43), after summation of the geometric series,

$$\begin{aligned}<\mathcal{T}> &= \mathcal{M}\left(\mathbf{cub}\left(0, \frac{(\rho-\gamma^*)}{1-\Xi(\rho-\gamma^*)}, \frac{(\gamma-\gamma^*)}{1-\Xi(\gamma-\gamma^*)}\right)\right) \\ &= \mathbf{iso}\left(0, \tfrac{1}{5}\left(\frac{2(\rho-\gamma^*)}{1-\Xi(\rho-\gamma^*)} + \frac{3(\gamma-\gamma^*)}{1-\Xi(\gamma-\gamma^*)}\right)\right) .\end{aligned} \tag{11.62}$$

Using the definition (11.42) with $\kappa^0 = \kappa$ and γ^0 corresponding to isotropic tensor $\mathbf{c}^*$,

$$\Xi = -\frac{2}{5}\frac{3\gamma^* + \kappa^*}{\gamma^*(2\gamma^* + \kappa^*)} \ .$$

By putting Ξ in Eq.(11.62) and setting $<\mathcal{T}>$ to be zero, we get the algebraic equation (11.53).

More general result can be obtained within terms of singular approximation. After substituting G by its local part $\mathcal{G}^0$ in the expression (11.61), one has

$$\mathcal{T} = (\mathbf{c} - \mathbf{c}^*)\sum_{i=1}^{\infty}\left(\mathcal{G}^0(\mathbf{c} - \mathbf{c}^*)\right)^i = (\mathbf{c} - \mathbf{c}^*)[\mathbf{I} - \mathcal{G}^0(\mathbf{c} - \mathbf{c}^*)]^{-1}$$
$$= [\mathbf{I} - (\mathbf{c} - \mathbf{c}^*)\mathcal{G}^0]^{-1}(\mathbf{c} - \mathbf{c}^*) = (\mathcal{G}^0)^{-1}\left((\mathcal{G}^0)^{-1} - \mathbf{c} + \mathbf{c}^*\right)^{-1}(\mathbf{c} - \mathbf{c}^*) \ .$$

(Shermergor and Patlazhan, 1976). Thus, the equation $<\mathcal{T}>= 0$ leads to

$$<\left((\mathcal{G}^0)^{-1} - \mathbf{c} + \mathbf{c}^*\right)^{-1}(\mathbf{c} - \mathbf{c}^*)>= 0 \ .$$

This is a form of the equation given originally by Kröner (1958). With the assumption of cubic crystal symmetry and strong isotropy, it is reduced to Eq.(11.53). ⊠

11.3 Related Issues

It is obvious that the theoretical considerations should be adequate to the real structure of the analyzed material. It is not easy to satisfy this condition, and the assumed model of polycrystals is a serious simplification of real materials. The presented methods are based on the assumption of perfect adhesion of crystallites. But this is not really true; the presence of boundaries leads to observable defect of effective moduli. The situation is complicated by the fact that the physical properties of a boundary depend on its numerous geometrical features and there are only vague ideas about the form of such dependence. Measured effective properties depend on grain size (e.g., Shirakawa and Numakura, 1958). Such a dependence is not admitted by any of the described methods; their results are scale independent, i.e., the calculated effective properties of two materials are equal if the microstructure of the first one is the same as the 'enlarged' microstructure of the second one. Other disregarded but essential factors influencing effective constants are the presence of defects of crystalline structure and residual stresses[5]. For more on the agreement of the calculated results and the experimental data see e.g., Kröner, 1986.

Closely related to the problem of polycrystal effective properties is the issue of determining single crystal properties from data for polycrystalline material. The problem is reverse to the determination of effective properties. Its solution could be helpful in getting crystalline moduli of materials difficult to obtain in the form of a monocrystal. The issue was approached by a number of authors, e.g., Druyvesteyn (1941), Gold (1950), Wright (1994). Full relation between poly- and single crystal constants involves correlation functions of all orders. Moreover, averaging causes loss of information and equations containing such procedures may have ambiguous solutions. Each of these factors indicates that the procedure of calculating single crystal constants can be only an approximation. The quality of such estimates remains to be investigated.

---

[5] It is worth noticing that for the influence of residual stresses on effective elastic constants, non-linear elastic properties of the material are a factor.

Another issue is getting information about microstructure of a material based on known elastic constants for mono- and poly- crystals. Obviously, the solution of such problem is also ambiguous and only crude approximations are possible. Practically, only the material's texture can be estimated but even in this case, strong additional assumptions are taken. Frequently, it is assumed that the constants of a polycrystal are given by the Voigt average (e.g., Sayers, 1982, Delsanto, Mignogna & Clark, 1987, Thompson, Lee & Smith, 1987). This estimation has relatively poor quality because only low (fourth) order coefficients of Fourier series of texture function are involved in the expression for the Voigt average, and only these coefficients can be calculated.

Let us finally mention that other internal characteristics of the material can be approached using methods described above. Having quantities necessary for calculation of $\mathbf{c}^*$ and using the relation $\varepsilon = \mathsf{A} <\varepsilon>$, expected values of quantities depending on ε or σ can be calculated. For example, the variance of the internal stress $<\sigma\sigma - <\sigma><\sigma>>$ can be obtained by using $\sigma = \mathbf{c}\mathsf{A} <\varepsilon>$. The averaging performed above over all orientations can be replaced by averaging over orientations satisfying a given condition. An important practical issue of that type is to get information on strain in crystallites of particular orientations. In the X-ray determination of internal stresses, one measures an average strain in crystallites with orientations O satisfying the relationship $\pm\mathbf{h} = O\mathbf{y}$, where $\mathbf{h}$ and $\mathbf{y}$ are unit vectors perpendicular to the diffraction plane in the crystal coordinate system and in the sample (laboratory) coordinate system, respectively (cf. chapter 10). I.e., the measured strain is equal to the average of $\varepsilon = \mathsf{A} <\varepsilon>$ over these particular orientations. This is the link between the macrovariable $<\varepsilon>$ and the investigated internal characteristic ε. It is crucial for the analysis of the internal stresses.

References

ALTMANN,S.L. (1986). *Rotations, Quaternions and Double Groups*, Clarendon Press, Oxford.

BARNET,D.M. (1972). The precise evaluation of derivatives of the anisotropic elastic Green's function, *Phys. Stat. Sol. (b)* **49**, 741–748.

BARTON,N.R. & DAWSON,P.R. (2001). A methodology for determining average lattice orientation and its application to the characterization of grain substructure, *Metall. Mater. Trans. A* **32**, 1967–1975.

BEATTY,M.F. (1966). Kinematics of finite, rigid–body displacements, *Amer. J. Phys.* **34**, 949–954.

BERAN,M.J. & McCOY,J.J. (1970). Mean field variations in a statistical sample of heterogeneous linearly elastic solids, *Int. J. Solids Structures* **6**, 1035–1054.

BERGER,M. (1987). *Geometry II*, Springer–Verlag, Berlin.

BHADESHIA,H.K.D.H. (1987). *Worked Examples in the Geometry of Crystals*, The Institute of Metals, London.

BIEDENHARN,L.C. & VAN DAM,H. (1965). *Quantum Theory of Angular Momentum*, Academic Press, New York.

BISHOP,J.F.W. & HILL,R. (1951). A theory of the plastic distortion of a polycrystalline aggregate under combined stress, *Phil. Mag.* **42**, 414-427.

BOLOTIN,V.V. & MOSKALENKO,V.N. (1968a). Macroscopic characteristics of microinhomogeneous solids, *Sov. Phys. – Dokl.* **13**, 73–75.

BOLOTIN,V.V. & MOSKALENKO,V.N. (1968b). The problem of determination of elastic constants of microinhomogeneouss medium, *Zh. Prikl. Mekh. Tech. Fiz.* (1) 66–72.

BOLOTIN,V.V. & MOSKALENKO,V.N. (1969). Towards calculation of macroscopic constants of strongly isotropic composite materials, *Mekh. Tverd. Tela* **3**, 106–111.

BONNET,R. (1980). Disorientation between two lattices, *Acta Cryst. A* **36**, 116–122.

BOOGAART,VAN DEN,K.G. (2002). *Statistics for Individual Crystallographic Orientation Measurements*, Shaker Verlag, Aachen.

BOTTEMA,O. & ROTH,B. (1979). *Theoretical Kinematics*, North–Holland Pub. Co., Amsterdam.

BROWN,C.M. (1989). Some computational properties of rotation representation, Technical Report 303, Univ. of Rochester Computer Science Dept.

BUCHAROVA,T.I. & SAVYOLOVA,T.I. (1993). Application of normal distributions on $SO(3)$ and S^n for orientation distribution function approximation, *Textures Microstruct.* **21**, 161–176.

BUNGE,H.J. (1970). Some applications of the Taylor theory of polycrystal plasticity, *Kristall und Techik* **5**, 145–175.

BUNGE,H.J. (1982). *Texture Analysis in Materials Science*, Butterworths, London.

BURENKOV,G.P. & POPOV,A.N. (1994). A method of automatically indexing Laue patterns, *Crystallog. Rep.* **39**, 556–561.

BUSING,W.R. & LEVY,H.A. (1967). Angle calculations for 3– and 4–circle X–ray and neutron diffractometers, *Acta Cryst.* **22**, 457–464.

CLÉMENT,A. (1982). Prediction of deformation texture using a physical principle of conservation, *Mat. Sci. Engng.* **55**, 203–210.

CORNWELL,J.F. (1984). *Group Theory in Physics*, Academic Press, London.

COXETER,H.S.M. (1946). Quaternions and reflections, *Amer. Math. Monthly* **53**, 136–146.

COXETER,H.S.M. (1973). *Regular Polytopes*, Dover Publishers, Inc., New York.

CRUICKSHANK,D.W.J. (1949). The accuracy of electron–density maps in x–ray analysis with special reference to dibenzyl, *Acta Cryst.* **2**, 65–82.

DEDERICHS,P.H. & ZELLER,R. *Elastische Konstanten von Vielkristallen*, Jül-887-FF, Jülich, 1972.

DEDERICHS,P.H. & ZELLER,R. (1973). Variational treatment of the elastic constants of disordered materials, *Z. Phys.* **259**, 103–116.

DELSANTO,P.P., MIGNOGNA,R.B. & CLARK,JR.,A.V. (1987). Ultrasonic texture analysis for polycrystalline aggregates of cubic materials displaying orthotropic symmetry, in *Nondestructive Characterization of Materials II*, ed. by J.F.Bussiere, J.P.Monchalin, C.O.Ruud & R.E.Green,Jr., pp.535–543, Plenum Press, New York.

DIAMOND,R. (1988). A note on the rotational superposition problem, *Acta Cryst. A* **44**, 211–216.

DIAMOND,R. (1990). Chirality in rotational superposition, *Acta Cryst. A* **46**, 423.

DOWNS,T.D. (1972). Orientation statistics, *Biometrika* **59**, 665–676.

DRUYVESTEYN,M.J. (1941). Die Elastische Anisotropie von Molybdän, *Physica* **8**, 439–448.

DYSON,D.J., KEOWN,S.R., RAYNOR,D. & WHITEMAN,J.A. (1966). The orientation relationship and growth direction of Mo_3C in ferrite, *Acta Metall.* **14**, 867–875.

EIMER,C. (1967). Stresses in multi–phase media, *Arch. Mech. Stos.* **19**, 521–536.

ENGØ,K. (2000). On the BCH–formula, *Reports in informatics*, Report no 201, University of Bergen, Bergen, Norway.

FAUGERAS,O.D. & HEBERT,M. (1983). A 3–D recognition and positioning algorithm using matching between primitive surfaces, *Proceedings of the Eighth International Joint Conference on Artificial Intelligence*, vol.2, edited by A.Bundy, pp.996–1002, William Kaufmann, Inc., Los Altos, California.

FOKIN,A.G. (1972). About applying the singular approximation to solving problems in statistical theory of elasticity, *Zh. Prikl. Mekh. Tech. Fiz.* (1)

98–102.

FORTES,M.A. (1983). Crystallographic applications of the elementary divisor theorem, *Acta Cryst.* **A39**, 348–350.

FORTES,M.A. (1984). Orientation relationships between two crystal lattices: matrix description, *Acta Cryst.* **A40**, 642–645.

FRANK,F.C. (1965). On Miller–Bravais indices and four dimensional vectors, *Acta Cryst.* **18**, 862–866.

FRANK,F.C. (1988). Orientation mapping, *Metall. Trans. A* **19**, 403–408.

GAIROLA,B.D.K & KRÖNER,E. (1981). A simple formula for calculating the bounds and the self–consistent value of the shear modulus of a polycrystalline aggregate of cubic crystals, *Int.J.Eng.Sci.* **19**, 865–869.

GANSTER,J. & GEISS,D. (1985). Polycrystalline simple average of mechanical properties in the general (triclinic) case, *Phys. Stat. Sol. (b)* **132**, 395–407.

GAZIS,D.C., TADJBAKHSH,I. & TOUPIN,R.A. (1963). The elastic tensor of given symmetry nearest to an anisotropic elastic tensor, *Acta Cryst.* **16**, 917–922.

GOLD,L. (1950). Evaluation of the stiffness coefficients for Beryllium from ultrasonic measurements in polycrystalline and single crystal specimens, *Phys. Rev.* **77**, 390–395.

GRIMMER,H. (1979a). Mathematical foundations for the description of the relative orientation of the grains in polycrystalline materials, abstract in *Symposium über mathematische Kristallographie*, Riederalp.

GRIMMER,H. (1979b). The distribution of disorientation angles if all relative orientations of neighbouring grains are equally probable, *Scripta. Metall.* **13**, 161–164.

GUBERNATIS,J.E. & KRUMHANSL,J.A. (1975). Macroscopic engineering properties of polycrystalline materials: Elastic properties, *J. Appl. Phys.* **46**, 1875-1883.

HALMOS,P.R. (1958). *Finite–Dimensional Vector Spaces*, Van Nostrand, New York.

HANDSCOMB,D.C. (1958). On the random disorientation of two cubes, *Canad. J. Math.* **10**, 85–88.

HASHIN,Z. & SHTRIKMAN,S. (1962). A variational approach to the theory of the elastic behaviour of polycrystals, *J. Mech. Phys. Solids* **10**, 343–352.

HEAD,A.K. (1979). The Galois unsolvability of the sextic equation of anisotropic elasticity, *J. Elasticity* **9**, 9–20.

HEAD,A.K. (1979). The monodromic Galois groups of the sextic equation of anisotropic elasticity, *J. Elasticity* **9**, 321–324.

HEINZ,A. & NEUMANN,P. (1991). Representation of orientation and disorientation data for cubic, hexagonal, tetragonal and orthorhombic crystals, *Acta Cryst. A* **47**, 780–789.

HERSHEY,A.V. (1954). The elasticity of an isotropic aggregate of anisotropic cubic crystals, *J. Appl. Mech.* **21**, 236–240.

HILL,R. (1952). The elastic behaviour of a cystalline aggregate, *Proc. Phys. Soc. A* **65**, 349-354.

HILL,R, (1963). New derivations of some elastic extremum principles, in *Progress in Applied Mechanics*, Prager Anniversary Volume, pp. 91-106, Macmillan, New York.

HIRSCH,J. & LÜCKE,K. (1988). Mechanism of deformation and development of rolling textures in polycrystalline FCC metals. I. Description of rolling texture development in homogeneous CuZn alloys, *Acta Metall.* **36**, 2863–2882.

HOFFMAN,D.K., RAFFENETTI,R.C. & RUEDENBERG,K. (1972). Generalization of Euler angles to N-dimensional orthogonal matrices, *J. Math. Phys.* **13**, 528–533.

HORN,B.K.P. (1987). Closed–form solution of absolute orientation using unit quaternions, *J. Opt. Soc. Am.* **A4**, 629–642.

HUMBERT,M., GEY,N., MULLER,J. & ESLING,C. (1996). Determination of a mean orientation from a cloud of orientations. Application to electron backscattering pattern measurements, *J. Appl. Cryst.* **29**, 662–666.

INOKUTI,Y., MAEDA,C. & ITO,Y. (1985). Transmission Kossel study of the formation on (110) (001) grains after an intermediate annealing in grain oriented silicon steel containing a small amount of Mo, *Metall. Trans. A* **16**, 1613–1623.

International Tables for X–Ray Crystallography (1952). Vol.1. Ed. by N.F.M.-Henry & K.Lonsdale, The Kynoch Press, Birmingham.

KABSCH,W. (1976). A solution for the best rotation to relate two sets of vectors, *Acta Cryst. A* **32**, 922–923.

KABSCH,W. (1977). A pattern-recognition procedure for scanning oscillation films, *J. Appl. Cryst.* **10**, 426–429.

KABSCH,W. (1978). A discussion of the solution for the best rotation to relate two sets of vectors, *Acta Cryst. A* **34**, 827–828.

KHATRI,C.G. AND MARDIA,K.V. (1977). *J. R. Stat. Soc. B* **39**, 95–106.

KLEIN,F. & SOMMERFELD,A. (1910). *Über die Theorie des Kreisels*, B.G.-Teubner, Leipzig.

KOCKS,U.F., TOMÉ,C.N. & WENK,H.–R. (1998). *Texture and Anisotropy*, Cambridge University Press, Cambridge.

KORN,G.A. & KORN,T.M. (1968). *Mathematical Handbook for Scientists and Engineers*, McGraw-Hill, New York.

KRAUSE,U., KUSKA,J.P. & WEDELL,R. (1989). Monovacancy formation energies in cubic crystals, *Phys. Stat. Sol. (b)* **151**, 479-494.

KRÖNER,E. (1953). Das Fundamentalintegral der anisotropen elastischen Differentialgleichungen, *Z. Phys.* **136**, 402–410.

KRÖNER,E. (1958). Berechnung der elastischen Konstanten des Vielkristalls aus den Konstanten des Einkristalls, *Z. Phys.* **151**, 504–518.

KRÖNER,E. (1967). Elastic moduli of perfectly disordered composite materials, *J. Mech. Phys. Sol.* **15**, 319–329.

KRÖNER,E. (1971). *Statistical Continuum Mechanics*, Springer-Verlag, Udine.

KRÖNER,E. & KOCH,H. (1976). Effective properties of disordered materials, *Sol. Mech. Arch.* **1**, 183–238.

KRÖNER,E. (1986). Statistical modelling, in *Modelling Small Deformations of Polycrystals*, ed. by J.Gittus & J.Zarka, p.229–291, Elsevier Applied Science Publishers Ltd, London.

KUIPERS,J.B. (1998). *Quaternions and Rotation Sequences*, Princeton University Press, Princeton, New Jersey.

LATTMAN,E.E. (1972). Optimal sampling of the rotation function, *Acta Cryst. B* **28**, 1065–1068.

LEIBFRIED,G. (1953). Versetzungen in anisotropem Material, *Z. Phys.* **135**, 23–43.

LIE,K.-H.C. & KOEHLER,J.S. (1968). The elastic stress field produced by a point force in a cubic crystal, *Adv. Phys.* **17**, 421-478.

LIFSHITZ,I.M. & ROZENTSVEIG,L.N. (1946). Towards the theory of elastic properties of polycrystals, *ZETF* **16**, 967–980.

LIFSHITZ,I.M. & ROZENTSVEIG,L.N. (1947). About construction of the Green's tensor for the fundamental equation of the elasticity theory in the case of unbounded elasto–anisotropic medium, *ZETF* **17**, 783–791.

MACKAY,A.L. (1977). The generalized inverse and inverse structure, *Acta Cryst. A* **33**, 212–215.

MACKENZIE,J.K. (1957). The estimation of an orientation relationship, *Acta Cryst.* **10**, 61–62.

MACKENZIE,J.K. (1958). Second paper on statistics associated with the random disorientation of cubes, *Biometrika* **45**, 229–240.

MACKENZIE,J.K. (1964). The distribution of rotation axes in random aggregate of cubic crystals, *Acta Metall.* **12**, 223–225.

MARDIA,K.V. (1972). *Statistics of Directional Data.* Academic Press, New York.

MARUTAKE,M. (1956). A calculation of physical constants of ceramic barium titanate, *J. Phys. Soc. Jap.* **11**, 807–814.

MATTHIES,S., VINEL,G.W. & HELMING,K. (1987). *Standard Distributions in Texture Analysis*, vol.1, Akademie–Verlag, Berlin.

McLACHLAN,A.D. (1972). A mathematical procedure for superimposing atomic coordinates of proteins, *Acta Cryst.* **A28**, 656–657.

MOAKHER,M. (2002). Means and averaging in the group of rotations, *SIAM J. Matrix Anal. Appl.* **24**, 1–16.

MORAWIEC,A. (1989). Calculation of polycrystal elastic constants from single-crystal data, *Phys. Stat. Sol. (b)* **154**, 535-541.

MORAWIEC,A. (1994). Review of deterministic methods of calculation of polycrystal elastic constants, *Textures Microstruct.* **22**, 139–167.

MORAWIEC,A. (1995). Misorientation-angle distribution of randomly oriented symmetric objects, *J. Appl. Cryst.* **28**, 289–293.

MORAWIEC,A. (1996a). Distribution of rotation axes for randomly oriented symmetric objects, *J. Appl. Cryst.* **29**, 164–169.

MORAWIEC,A. (1996b). The effective elastic constants of quasi-isotropic polycrystalline materials composed of cubic phases, *Phys. Stat. Sol. (a)* **155**, 353–364.

MORAWIEC,A. (1997). Distributions of misorientation angles and misorientation axes for crystallites with different symmetries, *Acta Cryst. A* **53**, 273–285.

MORAWIEC,A. (1998). A note on mean orientation, *J. Appl. Cryst.* **31**, 818–819.

MOSS,D.S. (1985). The symmetry of the rotation function, *Acta Cryst. A* **41**, 470–475.

MURA,T. & KINOSHITA,N. (1971). Green's functions for anisotropic elasticity, *Phys. Stat. Sol. (b)* **47**, 607–618.

MURNAGHAN,F.D. (1962). *The Unitary and Rotation Groups*, Spartan Books, Washington, DC.

NEUMANN,P. (1991). Representation of orientations of symmetrical objects by Rodrigues vectors, *Textures Microstruct.* **14-18**, 53–58.

NIGGLI,P. (1928). *Handbuch der experimental Physik* **VII**, Tail 1. Akad. Verlagsgesellschaft m.b.H, Leipzig.

NYE,J.F. *Physical Properties of Crystals, their Representation by Tensors and Matrices*, Clarendon Press, Oxford 1957.

PARDALOS,P.M., RENDL,F. & WOLKOWICZ,H. (1994). In *Quadratic Assignment Problem*, ed. by P.M.Pardalos & H.Wolkowicz, pp.1–42, American Mathematical Society, Providence.

PENROSE,R. (1955). A generalized inverse for matrices, *Proc. Cambridge Philos. Mag.* **51**, 406–413.

PIO,R.L. (1966). Euler angle transformations, *IEEE Trans. Automatic Control* **11**, 707–715.

PITSCH,W. (1959). The martensite transformation in thin foils of iron–nitrogen alloys, *Phil. Mag.* **4**, 577–584.

POSPIECH,J. (1972). Die Parameter der Drehung und die Orientierungsverteilungsfunction (OVF), *Kristall u. Technik* **7**, 1057–1072.

POSPIECH,J., GNATEK,A. & FICHTNER,K. (1974). Symmetry in the space of Euler angles, *Kristall u. Technik* **9**, 729–742.

PRENTICE,M.J. (1986). Orientation statistics without parametric assumptions, *J. R. Statist. Soc. B* **48**, 214–222.

RANDLE,V. (1996). *The Role of the Coincidence Site Lattice in Grain Boundary Engineering*, The Institute of Materials, Cambridge.

RANGARAJAN,A. CHUI,H. & BOOKSTEIN,F.L. (1997). The softassign Procrustes matching algorithm, in *Information Processing in Medical Imaging*, LNCS 1230, pp. 29–42, Springer-Verlag, Berlin.

RAO,S.N., JIH,J–H. & HARTSUCK,J.A. (1980). Rotation-function space groups, *Acta Cryst. A* **36**, 878–884.

REUSS,A. (1929). Berechnung der Fließgrenze von Michkristallen auf Grund der Plastizitätsbedigung für Einkristalle, *Z. Angew. Math. Mech.* **9**, 49-58.

ROSSMANN,M.G. & BLOW,D.M. (1962). The detection of sub–units within the crystallographic asymmetric unit, *Acta Cryst.* **15**, 24–31.

RYCHLEWSKI,J. (1984). Zur Abschätzung der Anisotropie, *Z. Angew. Math. Mech.* **65**, 256-258.

RYDER,P.L. & PITSCH,W. (1968). On the accuracy of orientation determination by selected area electron diffraction, *Phil. Mag.* **18**, 807–816.

SADANAGA,R. & OHSUMI,K. (1979). Basic theorems of vector symmetry in crystallography, *Acta Cryst. A* **35**, 115–122.

SAVYOLOVA,T.I. (1984). Grain distribution functions with respect to orientations in polycrystals and their Gaussian approximations, *Zavodskaya Laboratoriya* **50**(5), 48–52.

SAVYOLOVA,T.I. (1993). Approximation of the pole figures and the orientation distribution of grains in polycrystalline samples by means of canonical normal distributions, *Textures Microstruct.* **22**, 17–27.

SAYERS,C.M. (1982). Ultrasonic velocities in anisotropic polycrystalline aggregates, *J. Phys. D* **15**, 2157–2167.

SCHAEBEN,H. (1994). *Discrete mathematische Methoden zur Berechnung und Interpretation von kristallographischen Orientirungsdichten*, DGM Informationsgesellschaft Verlag, Oberursel.

SCHAEBEN,H. (1996). Texture approximation or texture modelling with components represented by the von Mises–Fisher matrix distribution on $SO(3)$ anf the Bingham distribution on S_+^4, *J. Appl. Cryst.* **29**, 516–525.

SCHAEBEN,H. (1997). A simple standard orientation density function: the hyperspherical de la Vallee Poussin kernel, *Phys. Stat. Sol. (b)* **200**, 367–376.

SCHAUB,H. & JUNKINS,J.L. (1996). Stereographic orientation parameters for attitude dynamics: A generalization of the Rodrigues parameters, *J. Astronaut. Sci.* **44**, 1–19.

SCHERINGER,C. (1963). Last–squares refinement with the maximum number of parameters for structures containing rigid–body groups of atoms, *Acta Cryst.* **16**, 546–550.

SHERMERGOR,T.D. & PATLAZHAN,S.A. (1976). Elastic constants of quasi–isotropic polycrystals, *Phys. Stat. Sol. (a)* **38**, 375–381.

SHIRAKAWA,Y. & NUMAKURA,K. (1958). On Young's modulus and grain size in nickel–copper alloys, *Sci. Rep. Res. Inst. Tohoku Univ.* **10**, 110–119.

SHUKLA,M.M. & PADIAL,N.T. (1973). A calculation of the Debye characteristic temperature of cubic crystals, *Rev. Bras. Fis.* **3**, 39-45.

SPIVAK,M. (1965). *Calculus on Manifolds: A Modern Approach to Classical Theorems of Advanced Calculus*, Westview Press, Boulder, CO.

STEPHENS,M.A. (1979). Vector correlation, *Biometrika* **66**, 41–48.

STUELPNAGEL,J. (1964). On the parametrization of the three–dimensional rotation group, *SIAM Rev.* **6**, 422–430.

SUSSMAN,J.L., HOLBROOK,S.,R., CHURCH,G.M. & SUNG–HOU,K. (1977). A structure – factor least-squares refinement procedure for macromolecular structures using constrained and restrained parameters, *Acta Cryst.* **A33**, 800–804.

SUTTON,A.P. & BALLUFFI,R.W. (1995). *Interfaces in Crystalline Materials*, Clarendon Press, Oxford.

TALPE,J. (1967). Use of a classical rotation operator in quantum mechanics, *Amer. J. Phys.* **35**, 114–116.

THOMPSON,R.B., LEE,S.S. & SMITH,J.F. (1987). Relative anisotropies of plane waves and guided modes in thin orthorhombic plates: implications for texture characterization, *Ultrasonics* **25**, 133–137.

THOMPSON,W. (1948). Note on the integration of the equations of equilibrium of an elastic solid, *Cambridge and Dublin Math. J.* (1846) reprinted in *Math. Phys. Paper* **1**, 97–99.

TOLLIN,P., MAIN,P. & ROSSMANN,M.G. (1966). The symmetry of the rotation function, *Acta Cryst.* **20**, 404–407.

TOLLIN,P. & ROSSMANN,M.G. (1966). A description of various rotation function programs, *Acta Cryst.* **21**, 872–876.

TSIOTRAS,P., JUNKINS,J.L. & SCHAUB,H. (1997). Higher order Cayley transforms with applications to attitude representations, *J. Guidance, Control, and Dynamics* **20**, 528–536.

VOIGT,W. (1889). Ueber die Beziehung zwischen den beiden Elasticitäts-constanten isotroper Körper, *Ann. Phys.* **38**, 573–587.

VOIGT,W. (1928). *Lehrbuch der Kristallphysik*, Teubner Verlaggeselschaft, Stuttgart.

WAHBA,G., FARRELL,J.L. & STUELPNAGEL,J.C. (1966). Problem 65–1, A least squares estimate of satellite attitude, *SIAM Rev.* **8**, 384–386.

WALPOLE,L.J. (1966). On bounds for the overall elastic moduli of inhomogeneous systems, *J. Mech. Phys. Solids* **14**, 289–301.

WALPOLE,L.J. (1985). The stress–strain law of a textured aggregate of cubic crystals, *J. Mech. Phys. Solids* **33**, 363–370.

WARRINGTON,D.H. (1975). The Coincidence Site Lattice (CSL) and grain boundary dislocations for the hexagonal lattice, *J. Phys. (Paris)* **36** (Suppl.

to No. 10), C4-87–C4-95.

WHITTAKER,E.T. (1904). *A Treatise on the Analytical Dynamics of Particles and Rigid Bodies*, Cambridge University Press, Cambridge, (reprint, 1988).

WRIGHT,S.I. (1994). Estimation of single–crystal elastic constants from textured polycrystal measurements, *J. Appl. Cryst.* **27**, 794-801.

YEATES,T.O. (1993). The asymmetric regions of rotation functions between Patterson functions of arbitrarily high symmetry, *Acta Cryst. A* **49**, 138–141.

ZELLER,R. & DEDERICHS,P.H. (1973). Elastic constants of polycrystals, *Phys. Stat. Sol. (b)* **55**, 831–842.

ZHAO,J. & ADAMS,B.L. (1988). Definition of an asymmetric domain for intercrystalline misorientation in cubic materials in the space of Euler angles, *Acta Cryst. A* **44**, 326–336.

Index

Druck: Strauss Offsetdruck, Mörlenbach
Verarbeitung: Schäffer, Grünstadt